A Mathematician
Comes of Age

© *2012 by the Mathematical Association of America, Inc.*

Library of Congress Catalog Card Number 2011943144

Print edition ISBN: 978-0-88385-578-2

Electronic edition ISBN: 978-1-61444-511-1

Printed in the United States of America

Current Printing (last digit):
10 9 8 7 6 5 4 3 2 1

A Mathematician
Comes of Age

Steven G. Krantz

Washington University in St. Louis

Published and Distributed by
The Mathematical Association of America

SPECTRUM SERIES

The Spectrum Series of the Mathematical Association of America was so named to reflect its purpose: to publish a broad range of books including biographies, accessible expositions of old or new mathematical ideas, reprints and revisions of excellent out-of-print books, popular works, and other monographs of high interest that will appeal to a broad range of readers, including students and teachers of mathematics, mathematical amateurs, and researchers.

777 Mathematical Conversation Starters, by John de Pillis

99 Points of Intersection: Examples—Pictures—Proofs, by Hans Walser. Translated from the original German by Peter Hilton and Jean Pedersen

Aha Gotcha and Aha Insight, by Martin Gardner

All the Math That's Fit to Print, by Keith Devlin

Beautiful Mathematics, by Martin Erickson

Calculus Gems: Brief Lives and Memorable Mathematics, by George F. Simmons

Carl Friedrich Gauss: Titan of Science, by G. Waldo Dunnington, with additional material by Jeremy Gray and Fritz-Egbert Dohse

The Changing Space of Geometry, edited by Chris Pritchard

Circles: A Mathematical View, by Dan Pedoe

Complex Numbers and Geometry, by Liang-shin Hahn

Cryptology, by Albrecht Beutelspacher

The Early Mathematics of Leonhard Euler, by C. Edward Sandifer

The Edge of the Universe: Celebrating 10 Years of Math Horizons, edited by Deanna Haunsperger and Stephen Kennedy

Euler and Modern Science, edited by N. N. Bogolyubov, G. K. Mikhailov, and A. P. Yushkevich. Translated from Russian by Robert Burns.

Euler at 300: An Appreciation, edited by Robert E. Bradley, Lawrence A. D'Antonio, and C. Edward Sandifer

Expeditions in Mathematics, edited by Tatiana Shubin, David F. Hayes, and Gerald L. Alexanderson

Five Hundred Mathematical Challenges, by Edward J. Barbeau, Murray S. Klamkin, and William O. J. Moser

The Genius of Euler: Reflections on his Life and Work, edited by William Dunham

The Golden Section, by Hans Walser. Translated from the original German by Peter Hilton, with the assistance of Jean Pedersen.

The Harmony of the World: 75 Years of Mathematics Magazine, edited by Gerald L. Alexanderson with the assistance of Peter Ross

A Historian Looks Back: The Calculus as Algebra and Selected Writings, by Judith Grabiner

History of Mathematics: Highways and Byways, by Amy Dahan-Dalmédico and Jeanne Peiffer, translated by Sanford Segal

How Euler Did It, by C. Edward Sandifer

Is Mathematics Inevitable? A Miscellany, edited by Underwood Dudley

I Want to Be a Mathematician, by Paul R. Halmos

Journey into Geometries, by Marta Sved

JULIA: a life in mathematics, by Constance Reid

The Lighter Side of Mathematics: Proceedings of the Eugène Strens Memorial Conference on Recreational Mathematics & Its History, edited by Richard K. Guy and Robert E. Woodrow

Lure of the Integers, by Joe Roberts

Magic Numbers of the Professor, by Owen O'Shea and Underwood Dudley

Magic Tricks, Card Shuffling, and Dynamic Computer Memories: The Mathematics of the Perfect Shuffle, by S. Brent Morris

Martin Gardner's Mathematical Games: The entire collection of his Scientific American columns

The Math Chat Book, by Frank Morgan

Mathematical Adventures for Students and Amateurs, edited by David Hayes and Tatiana Shubin. With the assistance of Gerald L. Alexanderson and Peter Ross

Mathematical Apocrypha, by Steven G. Krantz

Mathematical Apocrypha Redux, by Steven G. Krantz

Mathematical Carnival, by Martin Gardner

Mathematical Circles Vol I: In Mathematical Circles Quadrants I, II, III, IV, by Howard W. Eves

Mathematical Circles Vol II: Mathematical Circles Revisited and Mathematical Circles Squared, by Howard W. Eves

Mathematical Circles Vol III: Mathematical Circles Adieu and Return to Mathematical Circles, by Howard W. Eves

Mathematical Circus, by Martin Gardner

Mathematical Cranks, by Underwood Dudley

Mathematical Evolutions, edited by Abe Shenitzer and John Stillwell

Mathematical Fallacies, Flaws, and Flimflam, by Edward J. Barbeau

Mathematical Magic Show, by Martin Gardner

MAA Service Center
P.O. Box 91112
Washington, DC 20090-1112
800-331-1622   FAX 301-206-9789

For Charles Paine,
a good friend and a great teacher.
He was the first to give me a
glimpse of mathematical maturity.

Contents

Preface

The process of learning mathematics, more precisely of learning to be a mathematician, is a long and exacting one. It requires special discipline, and peculiar determination. It also requires a certain level of intelligence, but we will see in the discussions of the present book that intelligence is *not* the primary or determining factor.

As with any serious task, one should not embark on it unless one fully realizes what one is getting into. The purpose of this book is to explore what the task entails, who should engage in it, and what the rewards are.

The centerpiece of the mathematical education of any student is the intellectual development of that student. In grade school the child learns arithmetic and other basic mathematical operations. In middle school and high school there begins an exposure to algebra and other more abstract mathematical ideas. Geometry, trigonometry, and the theory of functions (that is, what is a function, and what does it do, and how do we manipulate functions?) follow in good order.

In today's world, however, the American K–12 student passing through this standard curriculum gets little or no exposure to rigor or to the concept of proof. Sophisticated problem-solving and analytical skills are not developed. As a result, the students that we have in our freshman calculus classes do not know what a proof is or what serious problem-solving is. Their problem-solving skills are nascent at best. They have rarely seen a proof, and are not equipped to create one.

If a student is to be a mathematics major and to become a practicing mathematician, then that student must become familiar with the traditional notions of mathematical rigor. This demand means that the tyro must learn about logic, set theory, axiomatics, the construction of the number systems, and proofs. The student must be able to read and evaluate proofs, but also must move on to being able to *create* proofs.

It is a considerable leap to develop from the textbook problem-solving state of mind so typical of lower-division mathematics courses to the theo-

retical, analytical, definition-theorem-proof state of mind that is typical of
real analysis and abstract algebra (and beyond). Many colleges and universities now have a *transitions course* to help effect this intellectual change
(see [DAW] for one of the most innovative books for such a course, and
[KRA3] for another). Teaching such a course is both a pleasure and a challenge; for one must determine how to get students over this hump. How
does one teach a student to put down the old iPhone and think hard about a
nontrivial mathematical proof?

The milieu described in the last paragraph raises the question of *mathematical maturity.* Every mathematician grows up hearing, at least in conversation, about mathematical maturity. It does not appear in the dictionary, or
in a standard reference work. But it exists. It is an idea that most mathematicians accept and profess to understand. We frequently make statements like,
"My course on elliptic functions requires a certain degree of mathematical
maturity." or "The goal of our mathematics major is to develop students
who have some mathematical maturity." We know what we mean when we
say these things, but we would be hard put to define our terms precisely.

It is curious that other disciplines do not speak of "maturity" as we do.
One does not hear of "English literature maturity," or "chemistry maturity,"
or even of "physics maturity". Other disciplines do not have the strict ver-
tical structure of mathematics (perhaps "tree-like structure" would be more
accurate), so their values and their vocabulary are bound to be different. In
history, literature, and philosophy there is nothing to prove. In chemistry,
physics, and biology, the exernal world is the arbiter of truth. In mathematics we must rely on our own minds to determine the truth. There is no other
judge. That is what sets mathematics apart.

Mathematical maturity consists of the ability to:

- handle increasingly abstract ideas
- generalize from specific examples to broad concepts
- work out concrete examples
- master mathematical notation
- communicate mathematical concepts
- formulate problems and reduce difficult problems to simpler ones
- analyze what is required to solve a problem
- recognize a valid proof and detect incorrect reasoning
- recognize mathematical patterns
- work with analytical, algebraic, and geometrical concepts

- move from the intuitive to the rigorous
- learn from mistakes
- construct proofs, often by pursuing incorrect paths and adjusting the plan of attack accordingly
- use approximate truths to find a path to a genuine truth

It requires a talented teacher, and also a good deal of drive and discipline on the part of the student, to achieve the goal of attaining mathematical maturity. How can we inspire our students to follow this path (the path that, presumably, we have all successfully followed)? How can we write texts and construct courses that will facilitate this transition?

These are the questions that we intend to tackle in the present book: What is mathematical maturity? How does the idea of mathematical maturity set us apart from other intellectual disciplines? How can we identify students who have mathematical maturity, or who are achieving mathematical maturity? How can we aid in the process?

You cannot succeed at anything in life unless you know what it is you are trying to achieve. In the case of being a successful and effective mathematics instructor, the nub of the matter is to help your students to achieve mathematical maturity. Any college teacher who takes the task seriously must address this issue, and find answers that work—consistent with that teacher's style of teaching.

This brief text will explore all aspects of the mathematical maturity question and endeavor to present some answers. A novice teacher will want to give careful thought to the issues presented here, and will want to internalize those issues as part of becoming an effective mathematics instructor. Senior faculty will also find ideas of interest here. Our presentation should cause even seasoned veterans to rethink what they do, and how they approach the teaching game. Resetting and adjusting our goals is part of development as a scholar.

A lot has been said about mathematical maturity in the context of informal coffee-table discussions. Not enough has been said about it in a rigorous, scholarly fashion. This book will be a first attempt to fill that void.

— SGK
St. Louis, Missouri

Acknowledgements

I would like to thank Harold Boas, John P. D'Angelo, Deborah K. Nelson, Robert Palais, Hung-Hsi Wu, and Doron Zeilberger for sharing their thoughts on mathematical maturity. Robert Burckel, David Collins, Roger Cooke, Jerry Folland, Jeremy Gray, Marvin J. Greenberg, Mark Saul, and James Walker read the entire manuscript with particular care, and contributed many incisive remarks.

As always, Don Albers was a supportive and vigorous editor who helped to bring the project along. He engaged an especially talented copy editor who helped me to sharpen and polish the prose. It is always a pleasure to work with Carol Baxter, Beverly Joy Ruedi and the other MAA book people. My thanks to all.

CHAPTER 1

Introductory Thoughts

Mathematical maturity is like pornography: I don't know what it is, but I know it when I see it.

John P. D'Angelo (mathematician)

A mathematician is mature up to the point where he becomes interested in mathematics.

Vladimir I. Arnol'd (mathematician)

I believe that a scientist looking at nonscientific problems is just as dumb as the next guy.

Richard Feynman (physicist)

Mathematics is the queen of the sciences.

Carl Friedrich Gauss (mathematician)

What I look forward to is continued immaturity followed by death.

Dave Barry (comic)

I, specifically, arrived at NYU intent on achieving a double major in film and . . . physics. Once I was informed of the required workload, especially the number of math classes I would have to take, I abandoned my scientific aspirations on the spot and focused my energy on filmmaking.

Nicolas Falacci (filmmaker)

1.0 Chapter Overview

What is mathematical maturity? How can we identify it? Perhaps more importantly, how can we recognize when it is not there, and then determine to do something about it? These are essential questions for any mathematics teacher, and ones that we must learn together to answer.

Mathematical maturity is elusive: We know what it is, but we do not know how to say what it is. It is important that we find explicit ways to describe it, to discuss it, and to come to terms with it. One of our goals is to recognize students who have the potential for mathematical maturity, and then to set them on the road to achieving it.

This chapter discusses the basics of mathematical maturity, its component parts, and its meaning. It will set one on the road to understanding mathematical maturity and learning how to nurture it.

1.1   Back Story

All practicing mathematicians think of themselves as having mathematical maturity. We do not wake up each morning and pay homage to the fact. We are not quite sure how to think about it consciously, but it gives us comfort.

Mathematical maturity is a developed sense of what mathematics is and how it works. Since mathematics has so many different manifestations and shapes, it may be difficult to imagine what that means. After all, there are those among us who formulate and prove theorems, there are those who develop algorithms, there are those who do mathematical modeling, there are those who calculate with *Mathematica*, and there are those who practice mathematical physics. Surely I am omitting or slighting several other groups. What do all these scientists have in common?

For one thing, they all have a robust and developed—and completely internalized—sense of mathematical logic. They all know how to read, analyze, and study rigorous argumentation. They all know how to *create* rigorous argumentation, and they all know how to record (for others to read and evaluate) rigorous argumentation.

Since the time of Euclid of Alexandria, we mathematicians have felt that the essence of who we are and what we do is to develop new results according to a universally recognized paradigm (see [KRA2] for the history and development of this point of view). We adhere to the axiomatic method, and follow strict (and universally accepted) rules of logic to move towards our mathematical goals. To be able to do this job well requires considerable training (although there are notable exceptions, such as Ramanujan), mental discipline, and skill. The agglomeration of these traits is what we think of as mathematical maturity.

It seems plain that a major goal of college or university mathematics teachers should be to instill in their students a sense of mathematical maturity. We are not simply trying to teach them formulas or rote techniques.

At least in upper-division courses, we are trying to teach them **(i)** mathematical technique, **(ii)** mathematical method, **(iii)** mathematical rigor, **(iv)** mathematical ideas. They have no hope of mastering any of these features of our subject without mathematical maturity.

So the mathematics instructor is faced with a considerable and very serious task: to figure out a way to develop mathematical maturity in students. This process is analogous to teaching your child how to be a good lover or a good friend. It is of immense importance, but it is highly nontrivial. It is something that one does largely for oneself, but it requires expert guidance.

I frequently teach a course (which I created) to acquaint students with set theory, logic, axiomatics, the construction of the number systems, and (most importantly) proofs. For the latter topic, students must learn to recognize, to understand, and to analyze proofs. They must learn to *create* proofs, and to tell when the proofs that they have created are in fact valid. This development is a considerable and delicate road, and I have spent many years finding how to help students over the pitfalls. I have written a textbook [KRA3] for the course, and it helps a lot. We have a lot of discussion in class, and I am often willing to take a good deal of time to go over relatively small points. It is a pleasure to see the students rise to the challenge, and take an interest in the material. At the end of the course, the students generally feel that they have learned something substantial, and have reached a new level of sophistication.

The course described in the last paragraph was created to bridge the gap between the relatively rote learning in lower-division mathematics courses and the more sophisticated, idea-driven learning in upper-division mathematics courses. Such a course is often, in the modern parlance, called a "transitions course." Before my course existed, instructors in real analysis and abstract algebra (and other upper-division courses too) would spend the first few weeks of class going over—guess what?—set theory, logic, axiomatics, and so forth. This paradigm seemed to me to be redundant and a poor use of time. Hence we have the transitions course which we now require of all majors (and which attracts many other students as well).

There are no doubt other ways to lead students along the path to mathematical maturity. Many schools have transitions courses. I had such a course when I was a freshman in college, and I can say honestly that it changed my life. The teacher was inspiring, the material was exciting, and the entire experience nailed down firmly my desire to be a mathematician. I can only hope that today's students can have a similar experience with the transitions course.

But I can also imagine that some schools would (instead of a transitions course) have tutorials, some would have workshops, some might even have labs. The purpose of this book is to analyze the concept of mathematical maturity, and to consider ways to develop it. [See [MOU] for some thoughts about mathematical maturity.] Let us begin.

1.2   First Principles

Where do we start? What are the grounding principles of mathematical maturity? I have always felt that first-order logic is the benchmark for analytical thinking, and that is where I like to begin.

When I was in ninth grade, it was an exciting time to be around Stanford University (where I grew up). The university was requiring all freshmen to take a course in first-order logic. The school system in Redwood City (ten miles up the road from Stanford, where I made my home) was in thrall to the great university, and was often inspired by what went on there. It created a course for gifted students in first order logic. I was lucky enough to be able to take such a course (actually taught to me privately by a Stanford professor [Neal Binford]—so I was even luckier than the other students).

This experience was my first introduction to genuine mathematical abstraction, and I loved it. It was clear to me that this was how I wanted to spend my life. The trouble is that I was one of the anointed. I was always meant to be a mathematician. Most of our students do not fit this mold.[1]

To repeat, most of our students—even the students in fairly advanced courses—are not necessarily future mathematicians. When I teach measure theory these days, most of the students are from finance (thanks to the Black/Scholes theory [BLS]). An abstract algebra course these days could easily have many future computer scientists and cryptographers. A differential geometry course could have students interested in gene folding and computer graphics. A dynamical systems course could have engineers, social scientists, medical researchers, or physicists. A graph theory course could have students from most any discipline.

---

[1] This insight is an important point, and I think we have the Teaching Reform movement to thank for it. For a long time we taught our elementary math courses in the French style—as though the students were all little mathematicians. In the old days, when it was customary for the top universities to flunk out half the freshman class, this custom was probably OK. It is no longer OK. In fact it is anathema. Our job now is to make it possible for all students to progress through the system. The vast majority of the students in our lower-division classes are *not* future mathematicians, nor anything like it. Some are future engineers or physicists. But many will be biologists, lawyers, physicians, geneticists, or business people. So we must formulate and communicate our ideas in a fashion that will be meaningful to such a diverse audience. We must learn anew how to teach.

When you teach a course now, you should try to get a sense of who the students are and what they are after. Let me stress that I am *not* advising you to pander. This is your course and you should teach it as you feel it. You are teaching at a good university, and one of the privileges of that rank is that you get to shape the curriculum as you see fit. But it can only improve your teaching, and its effectiveness, to know who is in your audience and what they are trying to achieve.

One of the most important attributes of good teachers is that they *listen*. You listen in a local sense to know whether students are following what you are talking about on any given day. But you also listen in a more global sense to learn something about your students—Who are they? What are their backgrounds? Why are they taking this course? What are their goals? The students can tell that you are listening, and they will respond in kind. You will have a more effective course, and a happier teaching experience.

This section is about guiding principles, and I have indicated that a good place to begin is with sentential logic. That is not the only possible jump-off point, but it is a good one. You could also begin with graph theory, or semigroups, or counting problems. The main point is to get the students thinking in new ways.

To look at this matter from a slightly different perspective: One of the bugaboos of modern mathematics teaching is that high schools teach calculus. When I was a student this early exposure to calculus occurred, but it was fairly rare. Now it is common.[2] Most of the students at my university have had some calculus in high school. The high schools teach calculus—so they tell me—because it is a good way for students to demonstrate that they can do college-level work.[3] But the process undercuts what the colleges and universities are supposed to be doing. We are the experts at teaching calculus, and it is pretty easy to argue that the world would be a better place if we were doing all the calculus teaching. But, for the most part, we are not.

---

[2]This is analogous to the fact that, when I was a child, very few children went to pre-school. There actually were very few pre-schools in those days. But, over time, parents convinced themselves that, if their kids did not go to pre-school, then they would be disadvantaged. So now most kids go to pre-school. Just so, today both parents and students (and teachers!) are convinced that if kids do not take calculus in high school then they will be "behind." One upshot is that most high schools teach calculus, and they make room for it in the curriculum by short-changing some of the essential topics in pre-calculus. Thus we have in our colleges over-educated students who do not know the basics—like how to add fractions.

[3]I was certainly interested when a group of high school teachers told me this. So I went to our admissions office and asked them whether it was true. They thought it was a fascinating idea, but they had never heard of it.

Unfortunately, many students who have had calculus in high school come to college and take calculus again. They say they are doing so because they don't think they got the real stuff in high school and they want to get it right. Many times colleges and universities have placement tests which only serve to show that the students' perception is correct: they did not get, or at least did not fully understand, the real stuff. They need to take calculus again. Some colleges and universities rely on the Advanced Placement tests to determine where students should be placed in calculus. Many times, independent of the placement exam, the student will repeat calculus because it is (apparently) easy and will not require much work. In any event, the entire situation makes freshman mathematics boring and repetitious.

Actually, the scenario is worse than what was described in the last paragraph. For freshman students are going to prioritize the classes that they have signed up for. If the student had calculus in high school and is now taking calculus again, then calculus is going to get shuffled to the bottom of the priority list. The student will not give the class adequate study or adequate effort, and will get a disappointing grade. It seems reasonably clear that we would all be better off if the high schools would do a better and more thorough job teaching the fundamentals (i.e., Euclidean geometry taught the right way with proofs, trigonometry, set theory, theory of functions) and leave it to us (at the colleges and universities) to teach calculus.

If we address the mathematical maturity question early on then we can begin to undercut some of this tedium. The topics that I mentioned earlier— first order logic, semigroups, graph theory, etc.—will definitely be new to students. These will provide a way to get the students to think about mathematics in new ways. The material will show them right off the bat that mathematics is *not* intrinsically boring and repetitious. It has a limitless variety of topics and ideas and is endlessly fascinating.

1.3   Next Steps

One is not born in a state of mathematical maturity. It is something one develops, just as one develops an ability to play the piano. The process requires drill and hard work and discipline. It does not just "happen." The notion that "Everything comes easy for him orher" is just nonsense. Mathematics does not come easy for anyone. It is just that some of us are more willing to do the necessary hard work than others.[4] Some of us have the dis-

---

[4]Malcolm Gladwell, in his fascinating book [GLA], asserts that it takes 10,000 hours of intense effort to master any subject area. Thomas Edison is famous for having said that "Genius is 1% inspiration and 99% perspiration."

cipline and tenacity, and others do not—see [DEV]. Some of us are lucky enough to have the right teachers and the right textbooks and the right environment that is conducive to study. Others are not. Some of us want this (i.e., mastery of mathematics) more than anything else. Others are less motivated.

It helps to have a peer group of equally fanatic mathematics students. There is hardly any better way to learn mathematics—or anything else—than by slugging it out as a group. As Uri Treisman [TRE] has shown us, the trick is to train students to work like professional mathematicians: They work for a while on their own, then they get together and exchange ideas and techniques, then they go back to working on their own. And so forth. It is a powerful yoga, and it works.

In fact one can go further in describing how a professional mathematician works. He or she will think about a problem, and then discuss it with a colleague. Then more thought is given to the problem, after which the mathematician may attend a seminar. Further thought is given to the problem, and then perhaps the mathematician again seeks out a colleague to have another chat. Certainly the mathematician will spend time reading, cogitating, and trying things. He/she will work out some examples, and try some experiments. Later on, perhaps the mathematician will give a seminar on partial results that have been attained. And so it goes. If things work out, eventually the mathematician will solve the problem—or at least make some substantial progress on it. It is this type of thinking, and this type of problem-solving strategy, that Treisman endeavors to teach to his students.

If you read the biography of a rock-and-roll star—see [RIC]—you will learn that the attitude of a budding rock star is that there is no other viable possibility. Either they will become a rock star, or else life is not worth living. For a mathematician it is the same. When I was young, I could not conceive of any other life to live but the mathematical life, and I believe that many of my colleagues felt the same way.[5] We cannot expect the students in our calculus courses, or even in our transitions courses, to possess this sort of fanaticism. But we should remember that, as teachers, one of our most important activities is being role models for our students. We can be the shining exemplars, and let them take from it what they will. I strive to show my students that I am thinking about mathematics most all the time, and that I have many interesting things to say about it. This includes

---

[5] There are exceptions. The great Hassler Whitney majored in music at Yale, and wanted to be a physicist. But he found that he could not keep track of all the facts in physics. His talents ended up being perfectly suited to mathematics, and we are fortunate for the career path that he chose.

information about the structure of the subject, about particular theorems, about mathematical history, about mathematical culture, about particular mathematicians, about mathematical politics, and many other features of the discipline as well. And I make it clear that living this way has given my life substance.

My students see that mathematics has given me a fulfilling and meaningful life, and I hope that they will take away the message that this way of life is a possibility for them as well. Mathematics today has many aspects and many dimensions. One can do mathematics by working on the genome project, or working as an actuary, or working at Los Alamos, or being a researcher at Lawrence Berkeley Labs, or being a professor, or working for Google, or for Hewlett-Packard, or for Mitre Corporation, or working at the Social Security Administration. There are an enormous number of possibilities.

1.4   The Mathematically Naive View of the World

Mathematically immature people have a Pavlovian or perhaps Skinnerian view of mathematics. Give such folks a solid of revolution and ask for its volume—they produce a formula and evaluate it. Or give them a matrix and ask for its eigenvalues—they draw out a dusty old procedure and crank through it. It would be easy to give dozens of examples.

By contrast, the mathematically mature person realizes that mathematics is about patterns, about how patterns develop and how they interact with each other. Such a person understands how specific tasks (as described in the last paragraph) fit into a broader, perhaps more abstract, picture.

There is a loose analogy with the game of chess. Chess novices learn a collection of openings and how they work. This practical knowledge will get them into the middle game. In the middle game, novices can only examine different possible moves and then select one that minimizes risk and maximizes the likelihood of victory. By contrast, chess masters can look at the board and discern broad patterns (covering up to five moves) and can thereby design a strategy. Of course a computer like Big Blue can look ahead ten moves or more and frequently beat even the most distinguished grand masters. But we would never say that Big Blue has mathematical maturity.

1.5 How Does Mathematics Differ from Other Fields?

We do not often hear about "history maturity," "literature maturity," "philosophy maturity," or even "chemistry maturity" or "physics maturity." Why is that? Is there a qualitative difference between those subjects and mathematics?

I would maintain that there is. There is a difference between a freshman history course and a senior or graduate history course. The history student must learn that history is not merely a collection of events and dates; it is in fact about *why* things happened and what effects they had. The latter requires greater sophistication and experience of the student. But the fact is that the skills for studying history are put in place early on. They develop and grow, but they are the same skills all along. It is the same for literature and philosophy and for almost any other subject. In freshman chemistry we are told that what chemists do is *measure things*. That statement is still true, to at least some extent, in a graduate chemistry course.

In philosophy and history there are no theorems. Much of the discourse is descriptive, and other parts are reactions to the theories of others. Literary critics do not work from definitions and axioms. They are analyzing and describing something much more poetic and fungible. It would be offensive to attempt a mathematical analysis in this context.

In physical science—chemistry, biology, or physics let us say—Mother Nature is the ultimate arbiter of what the truth is. Scientists can sit in their offices or labs and dream about how the world might work. But the final test is to look out the window and see how the world actually *does* work. One of the frustrating features of the exciting, modern subject of string theory is that there are no experiments to confirm any aspect of the subject. So far it is all conjecture and hypothesis. No experiments and no confirmations.

In mathematics the ultimate tester of our ideas is our own brains. Our brains enable us to formulate theorems and to *write down proofs*. These are proofs that other mathematicians can confirm and check and—ultimately—approve. This is the process by which mathematics is created and verified and put into the canon. It is quite distinct from the *modus operandi* of any other discipline. It is what sets us apart and makes us unique, and what guarantees that our ideas travel well and stand the test of time. Mathematical maturity is intimately bound up with the timelessness of mathematics.

Mathematics has a special nature. As the subject develops, so do the skill sets needed to master the new ideas. The techniques needed to learn calculus and solve calculus problems are different from the skills needed to learn abstract algebra and prove theorems about it. Likewise for real analysis, differential geometry, and other advanced mathematical topics. Freshman calculus involves basic problem-solving skills. Abstract algebra involves serious logic, axiomatics, set theory, sophisticated ability with proofs, and considerable tenacity. There is a marked qualitative difference between these two skill sets. The latter skill set is what mathematical maturity is all about.

1.6   A Little History

It was Plato (427 B.C.E.–347 B.C.E.) who taught us that there is a hidden geometry that governs the laws of nature. He believed that it was the so-called platonic solids—the tetrahedron, the hexahedron, the octahedron, the dodecahedron, and the icosahedron—and the way that an encompassing sphere would touch the vertices of these embedded polyhedra, that was the key to many physical phenomena. Plato used the platonic solids to represent atoms of the "four elements." He allowed that his system may not be the final word, but suggested that others could come up with better theories based on his.

Many years later, Tycho Brahe (1546–1601) and Johannes Kepler (1571–1630) formulated the idea that the way the Platonic solids sit in space governs the orbits of the planets. Kepler later abandoned this point of view, especially on account of his three laws of planetary motion, and developed a new way of looking at the motions of the planets, with a new geometry (based on the ideas of Appolonius regarding conic sections). Their ideas held great sway, and influenced Isaac Newton (1643–1727) to apply his calculus to confirm Kepler's Laws from first physical principles. See Section 1.12 for a more detailed analysis of the contributions of Tycho Brahe, Johannes Kepler, and also John Napier.

Independently, Galileo Galilei (1564–1642) was able to use his telescope to confirm and to challenge the ideas that were in the air about the shape of the universe.

In modern times, Albert Einstein (1879–1955) has caused us to rethink the geometry of the universe. For Einstein incorporated *non-Euclidian geometry* into his general relativity, and that idea has had a profound impact both on modern physics and on modern technology.

It is informative to ponder the evolution of mathematical maturity in the historical train of thought just described. By modern standards, Plato's geometry was naïve and simple-minded, but it was the cutting edge of thought in his day. After all, Euclid was setting the standard for modern mathematical axiomatic thought at roughly the same time that Plato lived. Kepler used ideas from the theory of conic sections to reformulate some of the long-standing ideas about planetary motion. Einstein was shrewd enough to take advantage of the relatively new ideas of Bernhard Riemann (1826–1866) about what we now call Riemannian geometry. He was able to conceive of the idea that space is curved, and that gravity is the cause of that curvature. This is a sophisticated (indeed, mature) idea, and one that Einstein himself did not fully understand. Modern workers on relativity theory—including Roger Penrose, Stephen Hawking, Edward Witten, and many others—have raised the level of mathematical sophistication in these studies to heretofore undreamed of plateaus.

In the past twenty-five years or so, yet a new level of recondite study has been achieved. For now we have string theory and superstring theory. This new set of ideas posits that the world is not composed of atoms and molecules—as we have believed for some time now—but rather of tiny strings that live either in 10-dimensional or 11-dimensional or 26-dimensional space (depending on what version of this lively theory you happen to subscribe to). In some ways string theory has now been superceded by M-theory and branes. It is difficult to keep up with all the developments. But the point is that string theory requires a new, very sophisticated geometry that is accessible only to an elite few. The level of maturity required to study string theory is formidable.

1.7 Examples of Mathematical Maturity

Mathematical maturity comes in many forms, but the unifying theme is an ability to deal with abstract (and often difficult) ideas. We never speak of mathematical *immaturity*—that would be politically incorrect, and in any event is a phenomenon that we prefer not to consider.

Grade-school students perhaps begin to exhibit mathematical maturity when they begin to ask questions. For this is when students start to realize that mathematics is not a fixed object, etched in stone. There are things that we do not know, and we do not understand. There are different possibilities to consider, and different directions that our studies may take.

It is when novice students ask *"Why is that true?"*—meaning that they want to see a demonstration of that fact—that the students are beginning to think like mathematicians. The novice is, in effect, asking for a proof. And a proof is a rather sophisticated entity, accessible only to a certain intellectual elite.

When a fourth-grader queries, "Why do we add fractions this way and not that way?" When a seventh grader asks, "Why can we do equal things to equal sides of an equation?" When a high school student ponders, "What happens to the trigonometric functions when we double the size of the triangle?" then he/she is beginning to exhibit mathematical maturity. When a college freshman says, "I don't know why the fundamental theorem of calculus should be true. It seems too good.", we know that this student is headed towards the good stuff. When a graduate student says to his/her thesis advisor, "I didn't like the proof that the author gave of this result, so I created my own proof," we know we are talking to a future mathematician.

It is well to recall the experience of I. I. Rabi (Nobel Laureate in Physics) as a child: When he was a boy returning home from school, his mother would usually say "Did you ask any good questions in school today?" What a perceptive mother! This is a wonderful way to gauge your child's progress in school.

Mathematics education professor Alan Schoenfeld characterizes mathematics as follows:

> Mathematics is an inherently social activity, in which a community of trained practitioners (mathematical scientists) engages in the science of patterns—systematic attempts, based on observation, study, and experimentation, to determine the nature or principles of regularities in systems defined axiomatically or theoretically ("pure mathematics") or models of systems abstracted from real world objects ("applied mathematics"). The tools of mathematics are abstraction, symbolic representation, and symbolic manipulation. However, being trained in the use of these tools no more means that one thinks mathematically than knowing how to use shop tools makes one a craftsman. Learning to think mathematically means (a) developing a mathematical point of view—valuing the processes of mathematization and abstraction and having the predilection to apply them, and (b) developing competence with the tools of the trade, and using those tools in the service of the goal of understanding structure— mathematical sense-making.

This is beautifully stated, and certainly summarizes the experience that many of us have had with the mathematical life. The social nature of mathematics is of pre-eminent importance, both for the functioning of the subject and for the mental health of its participants. It is a group effort. And, if one is going to master mathematics, then one needs to have a clear idea of what its tools are.

Let us conclude this section with an example of mathematical maturity that comes from the highest level. Around 1962, Fields Medalist John Milnor was giving a lecture in a large auditorium—to an audience of a couple of thousand people—about his celebrated theorem that there is more than one differentiable structure on the 7-sphere [MIL]. This was a *very* exciting result. If Milnor had instead discovered that there was more than one differentiable structure on the 1-sphere (the circle), people would not have been impressed. They would have said that this is just a remark, and anyone could have made it if they had only taken the trouble to think about the matter. But the 7-sphere is fairly exotic. And it turns out that Milnor's result is *not true* for the 1-, 2-, 3-, 4-, 5-, and 6- spheres. The 7-sphere is the first for which there is the phenomenon of more than one differentiable structure.

So Milnor is explaining this very sophisticated use of Stiefel-Whitney classes to a broad, undifferentiated audience of senior mathematicians. And, in the middle of the lecture, he clarifies an important point by saying, "This idea is best understood by looking at the 0-sphere." Is that not wonderful? The 0-sphere consists of only two points! What can one possibly say about the 0-sphere? Well, if you want to make a point about left-right symmetry, the 0-sphere may be the right way to do it.

It takes genuine mathematical maturity to be able to look at one of the most recondite parts of modern mathematics and elicit from it an idea that is so crystal clear and so incisive that one can explain it in terms of a simple idea like the 0-sphere. Even your grandmother would understand that explanation!

It is worth noting that John Milnor came to solve this problem because he was doing a calculation and things were not working out. There seemed to be some contradiction. Milnor banged his head against this thing for quite some time, with no result. Most of us would have long since given up— simply said that the problem was intractable and we could better spend our time in another pursuit. But John Milnor is no ordinary mathematician. He has brilliance, tenacity, and confidence. He knew that he could get to the bottom of the matter, and he did.

This is no isolated incident. Hans Lewy got confused over a problem of analytic continuation in complex 2-space and ended by discovering the Lewy locally unsolvable partial differential operator. Jean Esterle and Garth Dales became extremely frustrated over thirty years of efforts to prove that ring homomorphisms of Banach algebras are automatically continuous, and ended up producing a counterexample. Yum-Tong Siu became confused over calculations with the Kodaira Vanishing Theorem and ended up proving a new result about uniformization in several complex variables.

Surely this is high-level mathematical maturity at its finest. To face a confusing, even a contradictory, situation, and stick with it, and finally battle it to an enlightening conclusion, is the pinnacle of good mathematics. It is what we all strive for.

As a counterpoint, consider the experience of a friend of mine at Michigan State University. He was a precocious undergraduate (he won the Putnam Exam three times). He and another bright undergraduate were regular attendees at the weekly math colloquium. At one particular such event, the speaker was an algebraist whose topic was quasi-hemi-semi-demi Moufang loops. This fellow went to the blackboard and announced that he had invented a new algebraic structure that was quite interesting, and he wanted to present it and prove some results about it. He wrote down seven axioms for quasi-hemi-semi-demi Moufang loops and proceeded to state and prove theorems. About halfway through this performance, one of the bright, young undergraduates silently walked up to the blackboard and wrote down a three-line proof that the seven axioms were inconsistent. Quasi-hemi-semi-demi Moufang loops do not exist! Who in this melodrama is the mathematically mature one and who is not?

1.8   Mathematical Maturity and Fear of Failure

It has been suggested that mathematical maturity is the ability to sustain abstract reasoning over a long time span, and to do so without fear of failure.

This is perhaps putting the matter too strongly. We all fear failure. It is only human, and really smart people are perhaps more prone to fear of failure than others. People of modest intelligence fear failure too; perhaps they fail less frequently because they are better grounded.

An interesting analogy is with the world of sports. Consider the World Series in baseball. It consists of (at most) seven games. Imagine that each of the two teams has won three games. So they are down to the final game. Everything depends on the outcome of this game. And there are no draws in baseball. The two teams will fight this out to the bitter end.

The game would of course be televised, and viewed by many millions around the world. It would be the subject of analysis by hundreds of newscasters. Every move by every player would be put under the microscope. Every error would be analyzed. Every hit, every strikeout, every dropped ball, every fastball, would be subject to instant replay and intense scrutiny. How would the players feel in such a situation?

Clearly the players need to be able to concentrate on the game. If they spend too much time pondering the vicissitudes described in the last paragraph, then they will be distracted and will not be able to give the show their strongest effort. Players must treat each play as a stochastic process: the play that is happening right now does not depend on past plays and is

not predictive of future plays. They must give *this play* their best shot. Fear
of failure has no role in this scenario.

It was just the same in watching Jimmy Connors play tennis. He was
famous not only for his outstanding athletic prowess, but also for his temper
tantrums and ill humor on the court.[6] Why did he behave this way? It is
perfectly clear. Connors was completely absorbed in the game. For him,
nothing else existed in the world but *this point* in *this game* in *this match*.
His focus caused him to boil over with intensity and emotion. And hence
his somewhat childish behavior.

Just so with mathematical maturity. Mature mathematicians will get nothing out of moping over their past failures (and there will be plenty of them).
They should capitalize on years of mathematical experience—with some
successes and some misses. The main point is that mathematically mature
persons have been there and done that. They know what has to be done and
know how to do it.

Many research mathematicians have the same intensity of focus as did
Jimmy Connors. For them, there is nothing else in the world but the research
problem they are working on. You will see such a mathematician frequently
with a glazed look on the face, evidently unaware of the world out there.
Often such scholars are oblivious that people are talking to them, that it
is time to eat, that there is a storm brewing, or that life goes on. I have
even known mathematicians who have temper tantrums when frustrated by
a problem, or who will slam their heads against the wall in frustration and
aggravation.

Many of us will work on a problem for a year, look at the heap of calculations and attempts and missteps that we have accrued, and say "Well, this
has gotten me nowhere. I'll throw all this stuff in the trash and start anew on
something else." But really skillful mathematicians—those with mathematical maturity and the ability to apply it—will look at that pile of calculations
and be able to discern something worthwhile. They will say, "I didn't solve
the problem, but I learned something. Here it is, and I can write a nice little
paper about it." This distinction is a key to mathematical success.

One of the main points that we shall make in this book is that a major
step in the mathematical maturity process is to pass from calculations with
concrete numbers—such as

$$3^0 + 3^1 + 3^2 + 3^3 + 3^4 + 3^5 + 3^6 = 1093 \qquad (\star)$$

---

[6]Connors was once fined $15,000 for one of his outbursts. Rod Laver, a tennis star from
the old days, said, "We used to call that a pretty good year."

—to abstract thinking. It is a major epistemological step to see that the sum just indicated can be calculated with the formula

$$3^0 + 3^1 + 3^2 + 3^3 + 3^4 + 3^5 + 3^6 = \frac{3^7 - 1}{3 - 1} = 1093$$

or, more generally,

$$1 + \alpha + \alpha^2 + \cdots + \alpha^k = \frac{\alpha^{k+1} - 1}{\alpha - 1}$$

provided that $\alpha \neq 1$. It is the hallmark of a mathematically mature mind to look at a problem like $(\star)$—a problem which is ostensibly just an arithmetic problem—and say, "What is really going on here? Is there some general principle in play? Perhaps one could derive a formula. What could that formula be, and how might we discover it? Having discovered it, how could we then prove it?" You will *not* find this type of reasoning in the schoolyard. This is the epitome of mathematical reasoning at its best (at a very elementary level, of course). It is what a budding mathematician strives for.

When we get to the level of theorems and proofs in mathematics, then we are piling one piece of abstract reasoning on top of another. This is no longer a stochastic process (as with the baseball player discussed above). It is a very disciplined, and rather arduous, thinking process. Each step depends decisively on previous steps—and in a rather strict, prescribed fashion. It is a good first step to be able to look at an arithmetic problem like $(\star)$ and elicit from it an abstract mathematical principle. It is another decisive leap to develop from that stage to the level where one can formulate, understand, and begin to prove theorems. Many neophytes find the discipline too demanding. They simply do not have the tenacity (or the interest!) to fight through proof after proof. They move on to some less demanding field of study.

But mathematicians are *made* for this type of analysis. This is what they live for. This is what they seek. It is their avocation and their mantra. For a mathematician, there is nothing better than to create new mathematics and to *prove* that it is correct mathematics.

1.9   Levels of Mathematical Maturity

Mathematical maturity is not a single, well-defined state. A high school student is mathematically more mature than a grade school student. Likewise, a college student is more mature still, and a graduate student is actually (we hope) at a fairly sophisticated level of maturity.

A functioning faculty member at a good university—say an Associate Professor of Mathematics—has probably reached a certain pinnacle of maturity. Such a person is certainly capable of doing mathematical research. More, they can map out a research program, identify worthwhile problems, and develop means of attack. These faculty have probably had research grants, been invited to speak at conferences, and are recognized scholars.

By this measure, it can be said that I (the author of this book) am quite mathematically mature. I have written about 165 research papers and 65 books. I am a well-known and accomplished scholar. But the point that I want to make here is that there are people who are mathematically more mature than I. For example, the Fields Medalists are almost to a one more mature. Given any mathematical problem, or any mathematical situation, they can almost certainly handle it more adroitly than I, and with better effect. Whereas I might study the problem for a year and get some feeble partial results, they will quickly get incisive theorems that surely lead to further work and deeper insights.

The point is to see that mathematical maturity is a *process*. There is no pinnacle of maturity. If you are lucky, you continue to mature for your entire mathematical career. When celebrated cellist Pablo Casals was 85 years old he told an interviewer that he still practiced for at least two hours every day. The interviewer said, "But you are the greatest cellist who ever lived. Why do you need to practice?" Said Casals, "I can still see some improvement."

And that is how it should be. True scholars are never satisfied with where they have been or where they are going. Such persons always strive to do better. There are forever new challenges to face, new goals to conquer. It makes for a fulfilling life, one that we should all look forward to living.

To illustrate some of the points made in the last three paragraphs, I will relate a story about the distinguished geometer Blaine Lawson. A Professor at U. C. Berkeley and SUNY Stony Brook, Lawson is widely admired and respected. He was once giving a lecture on a problem he had been thinking about for twenty years. He had not solved it, but he made some interesting remarks about the nature of the problem, and offered some partial results. He noted that it was a *very* difficult problem, and he did not know when he would ever solve it. At the end of the talk, Fields Medalist Bill Thurston went up to Lawson and said, "Why don't you try this?" And it worked! Using Thurston's idea, Lawson solved his twenty-year-old problem.

In a slightly different vein, we might consider the experience of James Simons. In the late 1960s and early 1970s, Simons was a very accomplished differential geometer. He won the Veblen Prize, and his work with Chern is still cited today. Simons was also the founding Chairman of the mathematics department at SUNY Stony Brook, and deserves much of the credit for making Stony Brook the geometric powerhouse that it is today. To everyone's surprise, in the late 1970s Simons decided to quit mathematics. Cold. One reason he gave was that he had become interested in investing, and had decided to open an investment house. The other is that he had been working on a particular research problem for quite some time and it was driving him absolutely crazy. He could not face it anymore. So Simons opened Renaissance Technologies. This investment house was soon dramatically successful. Simons used mathematical ideas to develop investment strategies that were incisive. His hedge fund has been the most successful in history. In one year a few years ago Simons personally made more money than all the other mathematicians in the world combined. He is worth billions. Now Simons is retired, and is dedicating a substantial amount of his time to charitable works. Among other things, he has created the remarkable *Math for America* Program. And he is also thinking a bit about mathematics. One upshot of the latter is that he finally solved (in collaboration with Dennis Sullivan) the research problem that caused him to quit mathematics in the first place (see [SIS]). What a victory for mathematical maturity!!

1.10   The Changing Nature of Mathematical Maturity

Mathematics is immutable and unchanging. Mathematical facts—such as the Pythagorean theorem—that were established 2000 years ago are still valid (and useful) today. We can thank the Euclidean paradigm of axiomatic rigor for that stability and mobility of our subject. But the way that we see mathematics can change.

In the eighteenth and early nineteenth centuries, people believed that mathematics was a symbolic representation of facts about certain features of the real world around us. In other words, mathematics had foundations in reality. This is a very Platonic take on the subject.[7] Today, however, we acknowledge the Platonic parts of the subject but often focus on a more Kantian feature—that much of modern mathematics is quite abstract (i.e., is cooked up inside our brains) and has little or no basis in reality.

A mathematical neophyte—a child, for instance—will cling to the first point of view described in the last paragraph. Mathematical growth and maturation consists in significant part in transitioning to the second point

---

[7]More precisely, the Platonists believe that ideas live in a conceptual universe "out there somewhere."

of view. One purpose of this book is to discuss how that might be accomplished. The recent book [GRA] discusses this dialectic and what it means for mathematics.

The Kantian/Platonic dialectic is not a hot topic for debate among physical scientists. A chemist or a biologist would never claim that he or she had cooked up the latest theory strictly from his/her frontal lobes. Physical scientists are supposed to describe the world around us. Science is not just one more belief system—although some humanists have been known to claim such. Physical scientists are, by definition, Platonists. Put in other terms, nobody would be interested in a vaccine created by a priest.

There is considerable discussion these days of whether mathematics is Platonic or Kantian. The Platonic view is that mathematical facts have an independent existence out there in the ether, and we as mathematicians discover them. The Kantian view is that we mathematicians create mathematics in our minds. The reference [MAZB] has a nice consideration of the two points of view. As indicated three paragraphs ago, I would aver that we actually implement both points of view—and productively so. When we sit down to prove a theorem, we do not worry about whether the result is already floating out there in space or instead is being excreted from our pineal glands. Instead, we use any and all means to get our hands on the ideas and tame them to our purposes. There is no doubt that we just conjure up some definitions and concepts. We adapt them to the task at hand. Others are in the air—probably because someone else created them. It seems that there is a synergistic interrelationship between the Kantian and the Platonic, and we would do well to foster it.

1.11 On Proof and Progress in Mathematics

In 1994, William P. Thurston [THU] published a fascinating article with the same title as this section. He was not focused on mathematical maturity as such, but rather on higher-level issues (for a higher-level audience). Nonetheless, many of his remarks will resonate with readers of this book. We review a few of them here.

First of all, Thurston asserts, the main thing that mathematicians do is to advance human understanding of mathematics. Most of the ideas that he develops in [THU] are predicated on that hypothesis. Our main activity is *not* to prove theorems, it is *not* to win prizes, it is *not* to earn grants, it is *not* to get fancy job offers. The main goal is to increase the aggregate understanding of mathematics. This seems like a worthwhile avocation, and fits

in naturally with the idea of mathematical maturity. If we can raise the overall mathematical maturity of the human race, we have really accomplished
something.

Thurston considers that people have different ways of understanding a
mathematical concept. He illustrates this point by way of the derivative
(from calculus). Different ways of thinking about the derivative are

(1) **Geometric:** The derivative is the slope of a line tangent to the graph
of the function, if the graph has a tangent.

(2) **Rate of Change:** The derivative is the instantaneous speed of $f(t)$,
where t is time.

(3) **Microscopic:** The derivative of a function is the limit of what you get
by looking at its graph under a microscope of higher and higher power.

(4) **Infinitesimal:** The derivative is the ratio of the infinitesimal change in
the value of a function to the infinitesimal change in the variable.

(5) **Symbolic:** The derivative of x^n is nx^{n-1}, the derivative of $\sin(x)$ is
$\cos(x)$, the derivative of $f \circ g$ is $(f' \circ g) \cdot g'$, etc.

(6) **Logical:** We say that $f'(c) = \alpha$ if and only if for every $\epsilon > 0$ there is
a $\delta > 0$ such that, when $0 < |\Delta x| < \delta$, then

$$\left| \frac{f(c + \Delta x) - f(x)}{\Delta x} - \alpha \right| < \epsilon .$$

(7) **Approximation:** The derivative of a function is the best linear approximation to the function at a point.

We have modified Thurston's ordering here in an attempt to reflect in-
creasing sophistication, or maturity. In perhaps an attempt at levity, Thurston
concludes with an eighth conception of the freshman-calculus derivative:

(8) **Functorial:** The derivative of a real-valued function f in a domain D
is the Lagrangian section of the cotangent bundle $T^*(D)$ that gives the
connection form for the unique flat connection on the trivial $\mathbb{R}$-bundle
$D \times \mathbb{R}$ for which the graph of f is parallel.

There is a lot of beautiful mathematics in this last definition of the "deriva
tive" concept, but it is mathematics that most people would not understand.
The definition amounts to killing a flea with an atom bomb. We note, however, that working one's way through this list of definitions of a relatively
familiar concept is a small journey through mathematical maturity.

One of Thurston's most incisive ideas is that human thinking and understanding do not follow a single track. There are several different thinking faculties at play. Among these are:

(1) Human Language: "Our linguistic facility is an important tool for thinking, not just for communication" [THU, p. 164]. The language that we use reflects both the level and the profundity of our thinking. We learn the quadratic formula, as a simple example, almost as a chant. Thurston notes that freshman calculus students know only one mathematical verb: "equals." This is why students instinctively write such nonsense as $x^3 = 3x^2$ when what they mean to write (or say) is $\frac{d}{dx}(x^3) = 3x^2$.

(2) Vision, Spatial Sense, Kinesthetic (Motion) Sense: People, by nature, have a powerful intuition for assimilating visual and kinesthetic information. They are less well equipped for reversing the process—for turning internal insights into visual products. The scale of a visual structure can have a profound impact on this process: We are more comfortable with large-scale visuals.

(3) Logic and Deduction: We have built-in mechanisms for reasoning, and these tend to predominate over more formal logical processes. A mathematician solving a problem tends to think intuitively, and to skip over logical technicalities and difficulties.

(4) Intuition, Association, Metaphor: People have a terrific innate ability to sense an idea without knowing where it came from. This is intuition. Likewise, they can sense relationships between ideas without knowing exactly what the relationships are. It is important to learn to listen to one's intuitive perceptions. This is how deep ideas are developed.

(5) Stimulus-Response: This is learning to develop knee-jerk reactions to certain types of questions. If I ask you to calculate the eigenvalues of a matrix, you respond with a learned drill. If I ask you to find the tangent hyperplane to a surface in space, you trot out a standard procedure. We build our more sophisticated ideas on a rote collection of these processes.

(6) Process and Time: It is natural for us to think about processes or sequences of actions. We often couch more sophisticated ideas in that language. We think of a function as a process. Often a proof is perceived as a sequence of actions.

Mathematics has a very specific language—consisting of symbols, technical definitions, computations, logic, and other features as well. We use this language very economically to *record* our subject for the archive. But, in some ways, this language stands in the way of mathematical thought. When we are thinking about a problem, it is most effective to free oneself from the drudgery of notation and definitions and just let the concepts flow through one's cerebellum. When you read a mathematics paper—even a very technical one—understanding is best achieved by reading between the lines, by prying out the underlying ideas. Really great mathematicians will read a paper by just looking at the statements of the results; then they come up with their own proofs.

Many times a subject in mathematics will develop so rapidly that very little gets written down. Like a topic in ancient folklore, the collected wisdom is preserved by word of mouth. This can include lectures at conferences and colloquia, but it also includes many private conversations and even e-mail messages. Over time, people feel compelled to write books and papers just so that there is a formal record of what has been developed. While this last step is a good and valuable one, it also tends to stultify the subject, and to slow it down. Much of its vibrancy is squelched in this process.

The good news is that important ideas will be taught to the next generation of mathematicians, and they will reinvent the ideas and formulate them in their own language. In that way they bring the ideas back to life, and then new progress is made. This is exciting to witness, and it is what keeps us going.

The unifying theme of the last three paragraphs is that the important *lingua franca* in mathematics is *ideas*. We must all master the art of communicating ideas. Mathematical notation, as effective as it is for recording mathematics, is not good at getting ideas across.[8] One of the reasons that we all value colloquiua is that a good colloquium can convey ideas in ways that the written word cannot. In a single hour, good speakers can bring a whole group of people up to speed in a deep subject. Speakers can present complicated ideas in a plausible way, and help to make people comfortable with those ideas. They provide motivation for people to learn more, perhaps

---

[8]I have always marveled over musician Paul McCartney's claim that he cannot read music. And that is for the better, he asserts, because the ability to read music would interfere with his creativity. One cannot imagine mathematicians saying that they are glad to be unable to read mathematical notation, because doing so would interfere with their creativity. Mathematical notation is, in my view, an essential part of mathematical thought. One could not do mathematics without it. Even Ramanujan wrote things down, and used mathematical notation. Indeed, the Ramanujan notebooks are a valuable part of our mathematical heritage and literature.

to go and read the relevant papers. Also to have private conversations about the ideas on a later day. This is how, in practice, mathematics develops.

Mathematical theory tends to be developed and validated by consensus. In any given field of mathematics, a few of us battle our way through the proof of a tough new theorem. But those few then hold seminars and have private conversations and help bring others into the fold of those who believe in the new result and want to use it in their own work. It is a group effort, and a valuable one.

1.12   More on Brahe, Kepler, and Napier

In Section 1.6 we touched on Tycho Brahe and Johannes Kepler and their role in the development of our perception of the solar system. Let us now expand on those ideas.

Brahe was a noted astronomer in his day. He was fortunate to have a wealthy sponsor, who situated Brahe on his own island with an observatory. What a gig!

Brahe's student Johannes Kepler was also interested in the heavens, particularly in the motions of the planets. He was anxious to get his hands on Brahe's years of data, gathered from his detailed astronomical observations. But Brahe did not want to give up this data.

Now it is well known that, in those days, scientists in general were not prone to share their discoveries and their ideas. Sponsors and support were hard to come by, and there was much jealousy. After all, the first scientific journal did not come about until Henry Oldenburg founded *The Philosophical Transactions of the Royal Society of London* in 1665. Prior to that, Oldenburg had acted as a go-between for scientists, arranging a quid (such as the gift of a rare book) in exchange for a scientific secret. Such considerations were perhaps part of Brahe's reluctance to share his hard-won data with Kepler (both were strong-willed men), but perhaps an over-riding consideration was that Brahe feared that Kepler would use the data to validate the Copernican theory of planetary motion—that the planets orbited around the sun. Brahe himself subscribed to a modified version of the Ptolemaic theory—that the planets (except for Venus and Mercury) orbit around the Earth.

Now it turned out that Tycho Brahe had certain obligations to his sponsor, including attending periodic social events and agreeing to be "shown off" to the guests. At one such event, Tycho Brahe drank too much beer. His bladder burst and he died. This was a moment of opportunity for Kepler.

He was able to negotiate with Brahe's family and obtain the much-needed data. As a result, Kepler was able to do his now-famous calculations of the planetary orbits and, as a result, formulate his time-honored laws of planetary motion. Particulary, using the data on planet Mars, Kepler published his first and second laws (in Latin) in *Astronomia Nova* in 1609. This paper was translated into English by Donahue. Kepler generalized the first two laws to all planets, and also enunciated the Sun's position at a focus of the elliptical orbit, in *Epitome Astronomiae Copernicanae* in three papers published in 1618, 1620, and 1621. Parts of this work were translated into English by Wallis. Kepler's third law was published in *Harmonices Mundi* in 1619 and translated into English by Aiton et al.

For the record, we now state Kepler's laws:

I. The orbit of each planet is in the shape of an ellipse.

II. The orbit of a planet sweeps out area at a constant rate.

III. The square of the orbital period of a planet is directly proportional to the cube of the semi-major axis of its orbit.

Kepler's calculations were immense and tedious. He could have made good use of John Napier's (1550–1617) new theory of logarithms. But he refused to do so because he could not understand the derivation of this theory. So Kepler did all the work by hand. And it took many years.[9]

Note that this was at a time in mathematical history when even the operation of multiplication was considered to be quite abstruse. Only certain specialists in certain countries (such as Italy) were considered to be masters of the craft. So Kepler was carrying out a very eclectic and specialized project.

What is interesting here, from our point of view, is that Tycho Brahe's work exhibits no mathematical maturity. He was a gatherer of data. He worked long hard hours, and gathered information that nobody had ever had before, but he exhibited no insight into the nature of mathematics. Kepler, by contrast, showed real sophistication in the way that he took the reams of raw data that Brahe's work provided and turned them into concrete mathe-

---

[9]An interesting historical note is that, instead of logarithms, Kapler used the theory of *prosthaphairesis*. This is a floating-point technique based on the cosine function, rather than the exponential function. The nub of the theory is the formula

$$(\cos a) \cdot (\cos b) = \frac{1}{2} \left(\cos(a + b) + \cos(a - b) \right).$$

A moment's thought shows that this identity expresses a product as a sum. It is useful in saving some computational efforts, but neither as efficient nor as effective as logarithms.

matical formulas that *contribute decisively to our understanding*. This is a real epistemological leap.

But Kepler's mathematical maturity had the flaw that he could not understand John Napier's theory of logarithms. In this sense Napier was ahead of Kepler. If collaboration were more the norm in the seventeenth century, as it is today, then perhaps Napier and Kepler could have worked together, and could have produced much more scientific work more efficiently. It is fun to rewrite history, and to speculate on what might have happened.

CHAPTER

2
Math Concepts

If we desire to form individuals capable of inventive thought and of helping the society of tomorrow to achieve progress, then it is clear that an education which is an active discovery of reality is superior to one that consists merely in providing the young with ready-made wills to will with and ready-made truths to know with ...

Jean Piaget (philosopher and natural scientist)

My familiarity with various software programs is part of my intelligence if I have access to those tools.

David Perkins (Professor of Education, Harvard)

[Mathematical maturity is] fearlessness in the face of symbols: the ability to read and understand notation, to introduce clear and useful notation when appropriate (and not otherwise!), and a general facility of expression in the terse—but crisp and exact—language that mathematicians use to communicate ideas.

Larry Denenberg (computer scientist)

I think mathematical maturity is about the confidence to follow abstract reasoning, and the ability to sustain abstract thinking over a long time span.

Hung-Hsi Wu (mathematician)

I spent most of a lifetime trying to be a mathematician—and what did I learn? What does it take to be one? I think I know the answer: you have to be born right, you must continually strive to become perfect, you must love mathematics more than anything else, you must work at it hard and without stop, and you must never give up.

Paul Halmos (mathematician)

2.0   Chapter Overview

God is in the details. What sorts of problems can be used to ferret out mathematical maturity? What aspects of the mathematics curriculum are essential to mathematical maturity? What activities in the math department are dedicated to the development of mathematical maturity and which are not?

How can computers play a role in developing mathematical maturity? Are real analysis and abstract algebra and topology and geometry the be-all and end-all of mathematical maturity? Are there other aspects of the mathematical pie that can play a productive role here?

What parts of the basic calculus course can contribute to mathematical maturity? What about linear algebra? Differential equations? Where does mathematical maturity begin? Does it ever end?

These are the questions that we consider in Chapter 2.

2.1   Problems that Can Exhibit
Mathematical Maturity

I once gave the following problem to a group of students:

> Fix a triangle T in the plane. Show that there is a dilate δT, some $\delta > 0$ (that is, a triangle similar to T with side lengths δ times the side lengths of T) so that the area of δT equals the perimeter of δT.

I was astonished and pleased when one of the students quickly piped up, "Well, perimeter grows linearly and area grows quadratically, so the two must intersect at some point." And that solves the problem. In more detail, if P is the perimeter of T and A is the area of T then

$$\text{perimeter of } \delta T = \delta \cdot P \equiv P(\delta)$$
$$\text{and}$$
$$\text{area of } \delta T = \delta^2 \cdot A \equiv A(\delta).$$

If we graph $y = P(\delta)$ and $y = A(\delta)$ on the same set of axes, then one will be a line through the origin of positive slope and the other an upward opening parabola with vertex at the origin. These will intersect at the origin (which is of no interest) and at some point in the first quadrant. This last gives the dilate that we seek.

It exhibits some real sophistication to see through a problem like this— even though it is a very elementary problem. If one merely sits down and endeavors to calculate, one will be quickly frustrated. It requires an idea to get to the finish line.

Here is a second example:

Set your calculator in degree mode. Now calculate

$$y = f(x) = \tan(\cos(\sin(x)))$$

for any value of x whatsoever. The answer will be 0.01745 to five decimal places, no matter what the value of x.

What does this mean? Is the function f constant? Is there some other explanation?

I thank Sheldon Axler for this last problem. Surely the function is not constant, for

$$f'(x) = \sec^2(\cos(\sin(x))) \cdot (-\sin(\sin(x))) \cdot \cos(x) \cdot (\pi/180)^3 .$$

The first function factor here never vanishes, the last function factor only vanishes at points of the form $90° + k \cdot 180°$, so if f' were identically zero then it would have to be that $\sin(\sin(x)) \equiv 0$; and that certainly is not true.

Instead we may reason as follows:

- The function sine takes values between -1 and $+1$, hence $\sin(x)$ is between -1 degree and $+1$ degree.

- Then we are looking at $\cos(\sin(x))$, and that must be very nearly 1, since the argument of cosine is very nearly 0.

- Finally, we are looking at the tangent of an argument that is very nearly 1 degree. And the tangent of 1 degree is—guess what?—0.01745.

It takes some confidence, and some experience, to jump in and do this kind of reasoning. It is all very elementary, but not for the faint of heart.

As a final example, we give a brief description of a problem that comes from the complete works of G. H. Hardy (see [HAR, v. VII, p. 485]). The question is to find an approximation to the "large positive root" of the equation

$$e^{e^x} = 10^{10} x^{10} e^{10^{10} x^{10}} .$$

[We have striven here to duplicate the typesetting in Hardy's (1877–1947) original article exactly.]

Hardy's analysis is as follows: Take the natural logarithm of both sides to obtain

$$e^x = 23.0 \cdots + 10 \ln x + 10^{10} x^{10} = 10^{10} x^{10} (1 + \epsilon) .$$

Hardy notes that ϵ here is quite small (less than $40/10^{10}$ because $\ln x < x^{10}$). Thus

$$x = 23.0 + \cdots + 10 \ln x + \epsilon' ,$$

where ϵ' is positive and less than ϵ. This last equation enables us to find a sequence of lower and upper limits for x.

We shall not reproduce all the details of Hardy's analysis, but instead refer the reader to [HAR, v. VII]. Hardy concludes that x lies between 63 and 67. He notes that a closer approximation could be found "with a little trouble."

Hardy's concluding remarks are

I have purposely chosen a rather complicated equation of its type. The points to observe are (i) that the factor $10^{10}x^{10}$ proves to be of no importance whatever, and (ii) that it is futile to try to be very accurate in the early stages of the work. In all my inequalities I have left a good deal to spare, in order to work in round numbers as far as possible; and I have lost nothing by so doing. This is why examples of this sort are instructive, and teach a sense of proportion. The great weakness of boys confronted with a numerical problem is that they cannot see where accuracy is essential and where it is entirely useless.

To be sure, Hardy reveals here what a great teacher he was, and also what insight he had into our thinking processes. Mathematical maturity oozes from his analysis at the end. To understand so well the role of accuracy and inaccuracy is marvelous. Surely any student would learn a good deal from thinking about this problem.

I thank G. B. Folland for bringing this last example to my attention.

J. E. Littlewood (1887–1977) points out in [LIT] that questions involving large numbers can be grist for the development of mathematical insight. For example, imagine examing a page of a large, fine-print encyclopedia with 100 lines per page. One's goal is to close one's eyes and draw a line or curve from the top of the page to the bottom. What is the likelihood that this line passes always through the spaces between words, and never through the middle of a word? At first blush this seems like an intractable query. But one can figure that there is a five-to-one chance of succeeding in any given line, and then compute the likelihood for 100 lines. The answer turns out to be about one in 10^{70}.

In a similar vein, what is the likelihood that a monkey pounding away on a typewriter will produce—word-for-word—the play *Hamlet* by William Shakespeare? With a little thought, it is easy to see that the chances are 1 in 35^{27000}.

Sir James Jeans (1877–1946) is remembered for, among other things, calculating that, with each breath we take, it is more than even odds that we inhale some molecules from Julius Caesar's last breath.

2.2 Approximate Solutions

I think that many of us, when we are teaching numerical methods of integration in a second-term calculus course, fail to observe that the idea of an approximate answer is a profoundly new one.

Our students have been in school for thirteen years, and every math problem they have ever solved had a fixed, crisp, clean numerical answer. That answer was a real number, or a precise formula, or some other mathematical expression. The idea of being given an integral, such as

$$\int_0^1 e^{-x^2}\, dx\,,$$

and finding a numerical value that is accurate to three decimal places, is truly a foreign concept.[1]

So there are several things that need to be explained to the students:

- What is the meaning of an approximate solution?

- How does one measure how accurate an approximate solution is when one does not know the precise solution? How do we compare?

- Why is an approximate solution often adequate for the task at hand?

- How does one calculate an approximate solution?

- How does one calculate an approximate solution with a pre-specified degree of accuracy?

- What is accuracy to within k decimal places? What are significant digits?

Doing mathematics approximately is a whole new world, with a whole new set of values. It definitely requires mathematical maturity to be able to move beyond the method of solving problems precisely and to replace it with solving approximately. One has to be willing to reinvent what integration is, and how it works. One must learn the concept of an error term. One must learn how to deal with estimates (or *inequalities*). These are all quite new to the calculus student, and some of them (like estimation) are rather difficult. In fact it is safe to say that estimation, and the techniques adhering thereto, is a "big idea" of mathematics. It is at the heart of mathematical analysis. And it travels well.

---

[1] To the extent that students have thought about this matter at all, they probably believe that a mathematically accurate answer is one that is precise to eight decimal places—because that is what their calculator gives.

I was once approached by the CEO of a pharmaceutical machinist's shop and asked to calculate the volumes and surface areas of various pills (or tablets). Some of their shapes—like the right circular cylinder—are straight-forward to handle. Seventy-five years ago, that was the shape of most any pill. But modern marketing strategies today entail that each brand of pill have its own special "caplet" shape. And in fact, for practical reasons of drafting, the caplet shapes were specified for me as several arcs of circles (of different radii) pasted together. Calculating surface areas and volumes for such shapes is rather tricky.

The people who asked me to do this job were *not* mathematically ma-ture. They had only had a bit of high school mathematics. No calculus. The way they had been handling the task before I came along was a two-stage process. In the truly old days (before personal computers), a client would come to them with a sketch of a particular pill's shape and say, "We want machine dies for this pill in a size that will hold $550 \, \text{mm}^3$." The machin-ists would make the machine dies in five sizes—some obviously too small, some too large, and some in between. And then they would present the cus-tomer with all of these and say, "You choose." Clearly the mathematicians had not become involved yet, and the problem languished.

After personal computers became available, the machinists took a new view of the matter. They purchased a *very expensive* computer-aided design (CAD) system from McDonnell-Douglas. This software came with about 200 thick volumes of documentation, and required considerable computer power to run. If they wanted to calculate the volume of a certain caplet, they would sit down and draw a detailed picture of the caplet using the CAD system. Then they would use a Monte Carlo method to bounce a particle around inside the shape they had created and thereby calculate the volume.

This last is all fine and well, but it took several hours to draw each figure and then to perform the necessary calculation. The CEO of this company told me that what he wanted to be able to do, when he was on the phone with a customer from Germany, is to calculate the volume or surface area in a couple of seconds, just by pushing a few buttons. Then he could continue and develop his business transaction in real time.

And that is what I was able to do for them. I *am* mathematically ma-ture, and I could ferret out a path that could get them what they wanted. [Of course I used numerical integration techniques—or *approximation*—to carry out my task.] I wrote a piece of software for them—complete with graphics and a developed user interface—that became the definitive tool in the business. I charged them $15,000 for my work twenty years ago, and this was an order of magnitude cheaper than the cash that they paid for the McDonnell-Douglas CAD system.

This tale is another example of mathematical maturity winning the prize. I had a much more profound view of what mathematics could do than these machinists had. I saw further, and came up with the right solution.

2.3 Computers and Calculators

It is a sad fact of life that we in the academic profession did not consciously choose to admit pocket calculators into our classrooms. We did not hold faculty meetings and discuss the matter and come to a pedagogically cogent decision. Instead we were co-opted by the calculator manufacturers. What do I mean by this?

Twenty-five years ago, if you went to the annual AMS/MAA meeting in January and sought out the display area, then you would find Texas Instruments and Hewlett-Packard and Casio and all the other calculator manufacturers with booths promoting their wares. And they all had offers of this nature: If the instructor will adopt this calculator for his/her class then he/she will get a free one. Well, who could resist such a temptation? In all the world's great religions there is a vignette of the prophet being tempted by Mammon. And so it went in the mathematics religion. We agreed to use calculators in our classes, or in our math labs, because we fell for the promo.

Even for those of us who were not so prone to sins of the flesh, in many cases our students just started using calculators. From years of experience in high school, they could not imagine a mathematical world without one. Just as today, many adults say that they cannot think without the aid of their iPhones, so many students feel that they cannot reason without a calculator in hand.

Everyone knows that students come to college being quite versatile in using calculators, and they depend on them in an essential way to perform many elementary mathematical tasks. We used to complain that students were unable to add fractions. Now that list could be expanded considerably.

And in fact computers present an even more profound imbroglio. There are now available—*for free* on the Internet!—computer systems that can solve most elementary calculus and statistics and linear algebra problems. [There are also commercial products that one must actually purchase, such as *Mathematica*®, *Maple*®, and *MatLab*®. Some of these are rather pricey, and they all have specialized syntaxes that one must learn.] Three systems of note are *WolframAlpha*®, *Sage*®, and *Maxima*®. Let us discuss the first of these freeware products.

Go to the URL

$$\texttt{www.wolframalpha.com/}$$

You are greeted by a window asking you to type in your computational request. Remember that *WolframAlpha* has no specialized syntax. If you type in

```
Integrate x^3 - x
```

then the machine almost instantly spits out the answer

$$\int (x^3 - x)\, dx = \frac{x^4}{4} - \frac{x^2}{2} + \text{constant}.$$

But there is more! The screen also displays a graph of the function. It calculates alternative forms of the integral. And it calculates some definite integrals for you too, and relates them to area.

But it gets worse. In the upper righthand corner of the display screen is a button labeled `<show steps>`. If you click on that button, you will see that *WolframAlpha* shows every step of the calculation in horrendous detail. The software might be of limited utility if all it did was supply the answer. But, since it provides a complete solution, the student need look no further.

Now let us challenge *WolframAlpha* by asking it to calculate

$$\int e^{-x^2}\, dx.$$

We all know that the antiderivative of this integrand cannot be written down in closed form (see [RIT] or [ROS1], [ROS2] for a complete analysis of how to tell when an integral does or does not have this property). But *WolframAlpha* is no fool. It has to take a few moments this time to think things over, and then it yields the answer

$$\int e^{-x^2}\, dx = \left\{ \frac{1}{2}\sqrt{\pi}\, \text{erf}(x) \right\}.$$

One may argue that this is an unsatisfactory answer, since erf is *defined* in terms of the integral we are trying to evaluate. But we gave *WolframAlpha* an essentially impossible task and it came up with a correct answer. Users can find a book of tables and look up values of erf. Or else they could go to the remarkable NIST Digital Library of Mathematical Functions (DLMF) (`dlmf.nist.gov/`) for the needed numerical information.

In principle the student should be accustomed to the idea that some functions cannot be calculated by hand—numerical techniques, or at least a hand calculator, are needed to get reasonably accurate values. The functions sine, cosine, logarithm, and exponential are common examples of this phenomenon. Perhaps erf should be added to that collection. But erf will be much less familiar to the student.

Let us try to twist *WolframAlpha*'s arm to see whether we can get some more concrete information about this nasty integral. Type in

```
integrate e^{-x^2} from - 1 to + 1
```

Now *WolframAlpha* thinks for a while and ultimately gives the answer

$$\left\{ \sqrt{\pi}\,\mathrm{erf}(1) \right\} \, .$$

You can see the nasty erf function rearing its ugly head again.

It is clear that, to make the best possible use of *WolframAlpha*, one needs some mathematical maturity. Students probably would not know what to think if they were confronted with an answer involving erf. I note also that, when integrating e^{-x^2}, *WolframAlpha* does not offer a $<$show steps$>$ button.

The OnLine software packages *Sage* and *Maxima* also offer mathematical calculation environments similar in spirit to *WolframAlpha*. *Maxima* is particularly interesting because it is a direct descendant of the early computer algebra system *MACSYMA* that was developed at MIT in the 1960s. We shall not provide any further details here.

2.4   Proving Little Theorems, Proving Big Theorems

We all dream of proving the Riemann hypothesis or some other big result. Not only would there be the satisfaction of having made a major contribution to mathematics, but there would be a variety of encomia, great job offers, speaking engagements, prizes, and the like. This is precisely what Andrew Wiles achieved when he and Richard Taylor proved Fermat's last theorem, and it was all quite glorious.

But this is not what fate holds in store for most of us. We all dream of proving big theorems—indeed we all have a secret collection of them that we think about in our private moments. But then we have a more public collection of accessible theorems that are the ones that we actually talk about with our colleagues. These are the problems and theorems that we have a chance of making a dent in during the course of this lifetime, the ones that we write papers about and write grant proposals about and teach our students about. This is just the practical reality of life. If you are a medical researcher then you may dream of finding a cure for cancer, but there are more immediate and accessible problems on which you spend the bulk of your time. If you are a painter then you may dream of rendering a work with the impact and magnitude of Picasso's *Guérnica*. But meanwhile you have to put food on the table and you concentrate on more immediate gratifications—like painting portraits.

There is nothing wrong with proving little theorems. Little theorems combine in nice ways to make a whole that is greater than the sum of its parts. Many of my little theorems have served a useful role in pushing my subject forward. Sometimes another mathematician's big theorem was inspired in part by one of my little theorems. And we hope that the process of proving the little theorems gives us strength, and helps us to tool up for a bigger result. We might prove one little theorem each year, and a big theorem every ten years. That could be considered to be a productive mathematical career.

Part of mathematical maturity is being able to put this information into perspective. We all would like to set the world on fire every day, but it is not going to happen. We have to make peace among who we are and what we would like to be and what we can actually accomplish. This set of ideas relates to tenacity and delayed gratification, which we discuss in Section 5.14. You can only do what you can do.[2] One of the many reasons that we teach is that it is a constructive activity that makes us feel that we are contributing to the profession (even if we are not actually proving a great new theorem). This state perhaps borders more on emotional maturity than intellectual maturity; but if you want to survive in the mathematical game then you need to have this set of parameters under control as well.

Some mathematicians—like Niels Henrik Abel—are just bubbling over with ideas. There may be a nice, crisp new idea every few days. Some of these ideas will be big ones with many consequences. One could spend a year or two working out all the implications of the idea. Others are little ones that you could record in a two-page paper and then move on. It is fine, and it is healthy, to have a mix like this. One would be overwhelmed to have an earth-shaking idea every few days. Some variation in the tempo is soothing and cathartic. And it is only human.

And we must not give short shrift to the activity of working out examples. Most learning is performed *inductively*—starting with specifics and working up to general principles. Thus examples are how we jump-start the thinking process. A good example shows us what is really going on, and

---

[2] Although a good teacher will show students that they can do much more than was ever imagined.

allows us to start to conceptualize. It can lead to one or more important theorems.

By the same token, counterexamples can play a pivotal role in the development of a field. For they show where our thinking has been misguided, and cause us to re-evaluate our plan of attack. As an instance, the Weierstrass nowhere differentiable function caused some wholesale re-thinking of the basic principles of real analysis. The Peano space-filling curve was (and still is) a real eye-opener, and gave a genuine boost to the development of topology and geometric analysis. The proof that there is no elementary formula for the solution of fifth-degree polynomials put a halt to hundreds of years of mathematical research, and changed the direction of the subject.

2.5 Mistakes

It is a fact that mathematical writing is riddled with mistakes. This may seem odd, for we mathematicians pride ourselves on the precision and strict logical structure of our discourse. Everything is formulated according to established, time-tested rules, and nothing can pass muster (and be published!) unless it is dead on. But the paper [FEF], which won its author the Fields Medal, is full of mistakes. The paper [KRA4], which won its author (in fact the author of this book) the Chauvenet Prize, has some definite errors. The paper [CAR], which won its author the Abel Prize, contains some slips—even on the first page!

Professional mathematicians are accustomed to the notion that anything that they read will contain errors. And, as a survival mechanism, they develop automatic "error detection" techniques which help them to correct those errors on the fly so that they may understand what they are reading (or hearing) and forge ahead. This is an aspect of mathematical maturity, and one that readily distinguishes the accomplished mathematician from the tyro.

We who teach understand this insight instinctively. If you are giving a basic math lecture, and if you make mistakes—even very small ones—then the students will get lost. They have neither the experience nor the insight to quickly fill in gaps and then move on. Instead, they shift gears, stop listening, puzzle over the imprecision, and then give up and start thinking about what to have for lunch. The experienced mathematician, sitting in the same lecture, will simply say *sotto voce*, "Oh, of course he means thus and such." And then continue to listen to what is being said.

It clearly requires insight and experience and genuine mathematical ability to be able to fill in someone else's gaps (it is hard enough to fill in one's own!). And to do so with panache and flair. Someone who is struggling to follow the reasoning step-by-step is not going to be able to do it. Instead the situation calls for someone who is similar to a chess master. That is, it requires someone who can see ahead five or six moves, who knows where the line of reasoning is heading, who can see the forest for the trees. The student sees only the next fallen limb in the path, and likely as not trips over it. The professional sees the long-term goal, and nimbly walks around—or jumps over—that pesky limb.

One of the most famous mathematical mistakes of all time is the one made by Andrew Wiles in the course of his pursuit of the proof of Fermat's last theorem. Recall that Wiles announced his result at a conference at the Newton Institute at Cambridge University in June of 1993. By August of 1993, Nicholas Katz of Princeton and Luc Illusie of Paris had found an error in Wiles's argument. A fundamental error. Naturally Wiles was devastated. But he is no ordinary mortal, nor a wilting flower. He decided to roll up his sleeves and fight. It took him an entire year, but by September of 1994 Wiles and his student Richard Taylor discovered a way—not to *fix* the error but—to circumvent the error. Thus the story has a happy ending and Fermat's last theorem is indeed finally proved.

This is an instance, painted on a large canvas, of a mathematician recovering from a near-fatal error and producing excellent mathematics as a result. It can only be said that Andrew Wiles proved himself to be a matheamtician of the very first rank by finding the strength to fight through this adversity. He surely learned a lot from the process, but he must also have suffered a good deal of anxiety and self-doubt. This is an arduous process for any mathematician, and can sometimes take on magnificent proportions.

By contrast, we can relate the experience of Andy Gleason. In the 1950s, Gleason established a mighty reputation for himself by, along with Deane Montgomery, solving Hilbert's fifth problem. They proved that any locally Euclidean group is in fact a Lie group—a marvelous and powerful result. This triumph increased Gleason's already considerable confidence, and he proceeded to attack the Poincaré conjecture (which was only recently proved by Perelman using startling new techniques that were unavailable in Gleason's day). And he came up with a proof. Naturally the thing to do then was to go to Princeton, where all the experts were, and give a presentation. And so he did. Gleason stood up in front of Lefschetz and the other bigwigs and began by saying that his marvelous new proof of Poincaré relied on a very simple and elegant lemma and here it is. And he wrote down the lemma. Immediately Solomon Lefschetz raised his hand and said, "Yeah,

I once thought that I had proved that lemma. But it's false." Both Gleason and Lefschetz were mathematically mature—even to an extreme—but one trumped the other.

2.6 Mathematical Beauty

A mathematician can frequently be heard to say that "This theorem is beautiful" or "This proof is very pretty." What is meant by such a statement?

First, that the assertion (or theorem) is obviously true. Second, that it is compelling. Third, that its proof is enlightening and profound and perhaps enchanting. G. H. Hardy said that

> The mathematician's patterns, like the painter's or the poet's, must be *beautiful*; the ideas, like the colours or the words, must fit together in a harmonious way. Beauty is the first test: there is no permanent place in the world for ugly mathematics.

Part of developing in mathematics is to garner an appreciation for mathematical beauty. Mathematical beauty is an abstract concept—something which only someone who understands and appreciates mathematics can enjoy and internalize. When I teach my course on set theory and logic, some of the students (with little or no prodding from me) begin to express an appreciation for the aesthetic appeal of some of the proofs. This makes me happy, for it is clear that these are students that I am getting through to. Eventually comes an ability to *create* mathematical beauty. That is what the mathematical life is all about.

2.7 Modeling

Mathematical modeling is an old idea. Isaac Newton was perhaps the father of the modern concept of mathematical modeling. Certainly he was a master of finding mathematical paradigms for physical systems. Today this circle of ideas has reached soaring heights of sophistication. There are plenty of people who make their living by creating mathematical models. David Ruelle, among others, is particularly noted for finding a compelling and accurate mathematical model for turbulence (he also created the concept of "strange attractor").

It seems to me that mathematical modeling is the very height of mathematical sophistication. It requires not only a deep understanding of mathematical ideas, but also very special insights into the way that physics works. The expert modeler must not only see the math and the physics, but also how the two subjects interact. This is perhaps a new type of maturity beyond what is being discussed in the present book. It is not enough, to be an expert modeler, to have a textbook understanding of basic mathematical precepts. One must in fact understand how the ideas were created and why they were created in that particular way. One must be able to manipulate the concepts in sophisticated, high-level paths that will lead to new insights. One must be a complete master of two entire fields.

It gives one pause for thought to think of mathematical maturity in this new context. For it is a very refined context, and a demanding one. It puts new strictures on its practitioners, demands that they far exceed what many of us may realistically expect to be able to do. It sets a new standard for excellence, one that most of us can only dream of.

CHAPTER **3**

Teaching Techniques

Proofs should be communicated only by consenting adults in private.

Victor Klee (mathematician)

I was of course flabbergasted by the variety of generalisations that have blossomed in that particular garden!

Bryan Birch (mathematician)

Computers are incredibly fast, accurate, and stupid. Human beings are incredibly slow, inaccurate, and brilliant. Together they are powerful beyond imagination.

Leo Cherne (economist)

An individual understands a concept, skill, theory, or domain of knowledge to the extent that he or she can apply it appropriately in a new situation.

Howard Gardner (developmental psychologist)

To be mature means to face, and not evade, every fresh crisis that comes.

Fritz Kunkel (psychiatrist)

The stone age did not end because we ran out of stones.

anon.

3.0 Chapter Overview

It is natural to wonder which teaching techniques will contribute to mathematical maturity. Do the tenets of teaching reform have something to show us about the matter? What about OnLine learning?

These days capstone experiences are in vogue. At the end of a four-year undergraduate education, the capstone experience helps the student draw together material from different courses, and to see connections that were previously unnoticed. It seems likely that such a project could help to push the student along the maturity path.

There are many new approaches to math teaching, from those of Uri Treisman to those developed at Harvard as part of the reform movement. They are all worth considering as we work to help our students move ahead in the mathematical maturity cycle.

3.1   Teaching Reform

The teaching reform movement had its formal beginning in 1986, when Ronald Douglas of SUNY, Stony Brook and Stephen Maurer of the Sloan Foundation organized a small (twenty-five participant) meeting, at Tulane University, to lament the sorry state of lower-division mathematics education —particularly calculus teaching. Some of the early reports pursuant to Douglas's meeting, and the ideas that sprang therefrom, appear in [NCE], [DOU] and [STE1]. For a more recent assessment, see [STK], [LET], [TUC], [KLR], or [ROB]. Other points of view are offered in [WU1], [WU2].

Douglas is a great organizer, and he got people fired up with information that our attrition and dropout rate in calculus is embarrassingly high, that our failure rate is unacceptable, that our teaching is not optimal, and—yes— that *the lecture is dead.*

Douglas persuaded the federal government to be interested in the problem, and many different programs were set up at the national level to encourage mathematics faculty—at universities and colleges of all sorts—to re-think and re-invent their curricula. There are special programs sponsored by the National Science Foundation for work on

- calculus
- precalculus
- post-calculus
- high school curriculum
- instrumentation and laboratory improvement
- discrete mathematics
- advanced technology education
- undergraduate faculty enhancement
- teacher preparation

and many others as well.

When Persi Diaconis was asked about his MacArthur Prize (see [ALA]), he said, "...if somebody gives you a prize and says that you're terrific, well, that's nice. But if someone gives you a prize and says here is $200,000 to show you that we mean it, maybe they really mean it." So it is with teaching reform. Government funding has made the reform idea into a movement, and has made it necessary for the rest of us to take it seriously. It has enabled the investigators to buy equipment, to run teacher-training workshops, to hold conferences, to write books, and to get the word out. It has really had an effect. Like it or not, the Harvard calculus book had the biggest first year sales of any calculus book in modern history—by a factor of about four. It has had an impact, and has caused many of us to take a new look at calculus. That effect alone may justify saying that the money was well spent.

Let me not mince words about one key notion. The British educational system—especially as it is practiced at Cambridge and Oxford Universities—does a terrific job of teaching students the art of discourse. Students at these extraordinary institutions are assigned tutors. Students meet weekly with their tutor, who literally drills them in the arts of written and oral communication. Each student's weekly assignment is to write an essay on a topic chosen by their tutor. Students use all the resources of the university—lectures, other faculty, the library—in preparing the assigned piece. Then they must meet with their tutor and defend it. It is a wonderful regimen, and serves to keep language and dialogue alive.

We in the United States have not traditionally done as well at this sort of training. Instead, we take eighteen-year-old freshmen, many of whom have never articulated a thought more sophisticated than an order for a burger and fries shouted at the plastic effigy outside of the local Jack-in-the-Box®, and we suddenly expect them to formulate and deliver cogent sentences about the chain rule and linear independence. Talking mathematics is a high-level skill. We should not simply *expect* our freshmen and sophomores to be able to do it. Instead, we should *plan to train them to do it*. We should develop techniques for *teaching* our students to talk to us. Many of the teaching reform projects—notably the Harvard project (see [HAL])—have led the way in teaching students the art of communication, and particularly how to write.

The hidebound among us, those who are content with traditional teaching methods and who have little patience for the reform practices, are fond of suggesting that "money talks and baloney walks." If I were a cynic, I would recall the story about Willie Sutton—the famous twentieth-century bank robber. When asked how he got into robbing banks, Willie said, "Well, that's where the money is." I know many of the principals who are involved

in calculus reform. Surely they like receiving summer salaries and other pecuniary support no less than the rest of us. But these are people who are
sincerely committed to bettering undergraduate education in mathematics.
And they probably work a lot harder at it than the rest of us. Their conclusions are certainly worth our careful examination.

Debate about reform tends to be quite emotional. I believe that the vitu-
peration is due partly to the fact that people are worried about defending
their turf. Also, middle-aged math professors probably do not want to discard the teaching techniques that they have been using half their lives and
then retool. But I have also found (and this phenomenon often holds when
people disagree vehemently and hopelessly) that different participants in
the reform discussion actually have different concerns and goals.

- Some are concerned with teaching a well-defined body of mathematics
  to a given audience.

- Some wish to guarantee that a certain body of knowledge is delivered to
  the students at a certain rate. So that, by the end of four years in college,
  they will have reached a certain well-defined plateau.

- Others are interested in maintaining and promoting the students' self-
  esteem, and in empowering certain groups.

- Still others wish to cut the attrition rate, so that more students can advance to upper-division mathematics courses—no matter what the loss
  in content and curriculum.

I am not judging or ranking the goals just adumbrated. But, if one reviews
the somewhat alarming debates about mathematics teaching that have been
going on in California (see [AND], [JAC1], [JAC2], and [ROSG]), one sees
in the end that frequently the warring factions are discussing entirely different matters. It is essential that we communicate, and find grounds for
cooperation rather than for hostility (see particularly the article [ROI] by
Judith Roitman). In the end, we all want to learn how to teach our students
so that they learn what they need to learn and go on to success in their later
activities.

Franklin Roosevelt said that "Politics is the art of the possible." There is
wisdom in these words. Nobody is prepared to discard wholesale the traditional teaching techniques that we have used for so long, and likewise no
thinking person can ignore the new needs of our students, the new technologies that are available, and the new ideas about teaching that are being
developed. Teaching is a lifelong passion, and we spend our entire lives
developing and growing as educators. We continue to learn both from the

reform movement and from re-thinking traditional methods of learning and teaching. Our ideas will continue to evolve.

What does the teaching reform movement have to do with mathematical maturity? Will a student taught by reform methods develop more quickly along the path of mathematical maturity than one taught by traditional methods? Will reform techniques and reform tenets reinforce mathematical maturity or detract from it? Will they cause students to develop more quickly or more sluggishly? Will they light a fire under students and reinforce their desire to be mathematicians, or will they quench those fires?

There has not been much hard-hitting analysis of whether, in general, reform techniques yield better results than traditional techniques. Do reform students retain the material better? Do they tend to perform more effectively in advanced courses? Do they go on to be more successful scientists? Nobody knows.

Everyone has their own opinion and their own agenda. It seems to me that getting students involved in self-discovery and group learning can only enhance the mathematical maturation process. The students still need to do all the hard work to learn the mathematics and the mathematical method. But communication skills are an essential part of reasoning and problem-solving. To the extent that reform techniques contribute to better and more effective speaking and writing skills, they are more part of the solution than part of the problem.

Many of us older faculty would argue that the best way to inculcate mathematical maturity in students is by way of the traditional grind: You study the books, you do the problems, you take the exams, and you slowly develop. That is how we did it, so it must be the right way. Well, we are all comfortable with what we know. But I think we can learn from some of the new ideas, and enhance the values that we hold dear. There is always room for improvement. I have learned a great deal from the teaching reform movement, and my entire approach to the process has developed and evolved as a result. I am not by any means a reformer, but I am a friend of the reformers.

3.2 Math Maturity and Math Teachers

It is natural to ask what sort of mathematical maturity K–12 mathematics teachers should have. Mathematics teachers are, after all, role models for students. They personify for the students what a mathematician should be.

Mathematics teachers should come across as a master of mathematics (at *that* level). They should be able to field most questions that come up in class, and be able to answer them in a constructive and creative fashion. It is OK for a teacher to say occasionally, "Let me think about that one. I'll give you an answer next time." But that should be the exception rather than the rule.

Mathematics teachers are, after all, paragons for students of what mathematicians are. When students go on to the university, then they will be taught by *genuine, practicing mathematicians*. That is actually very exciting, and sets a pattern for the student's future life. But, in grade school and high school, teachers are the be-all and end-all of mathematics. They should be something of an authority figure.

Unfortunately it is often the case in practice that the teacher falls short. Many grade K–12 teachers are poorly trained and are certainly not masters of their subjects. Many of them can barely read the texts to the students. That is what passes for teaching in our schools. The reasons for this malaise are sociological and economic. We pay our teachers so poorly that the best people are not attracted to the profession. The new *Math for America* program, sponsored by Jim Simons and headed up by John Ewing, is designed to address this impasse. But there is a lot of work to be done.

3.3   Uri Treisman's Teaching Techniques

Some years ago, Uri Treisman was a postdoc at the University of California at Berkeley. He was assigned to teach minority students who, traditionally, had struggled with calculus.[1] Treisman took the task seriously. He made a list of all the reasons one might suppose that minority students would do poorly in such a course. On his list were:

- Poor home environment.

- Parents uneducated.

- No books at home.

- Students have outside jobs.

- Students have spouses and families.

- Students have a weak math background.

- Students live far from school (perhaps in a ghetto) and must ride the bus for hours to get to class.

There were many more. As a first step in determining how to deal with these students, Treisman investigated each of the items on his list to determine

---

[1] There were few African-American students at Berkeley in those days, and their failure rate in the basic calculus courses was *very* high.

which of them actually had a bearing on the students he was to teach. It turned out that *none* of them did.

There is an interesting epistemological point here. In high school, most learning takes place *in the classroom* at the moment of impact. The primary responsibility for learning lies with the teacher. In college, most learning takes place *outside the classroom*. The primary responsibility for learning lies with the student. This is a fundamental difference, and part of learning how to succeed in college is understanding this difference and *learning how to study*.[2]

One very important insight that Treisman had was this. He noticed that Oriental students were trained, from a young age, to work in groups. Thus they had a social support infrastructure to lean on when things got tough. By contrast, the African American and Latino students at Berkeley had mostly been the best students in their respective high schools, but they were loners. They had no social support infrastructure. When they got into calculus and hit the wall, they had nobody and nothing to lean on. This gave Treisman one of his key ideas for helping the students. Read on.

After Treisman worked with his particular group of minority students for a while, he determined that their main problem was that they did not know how to study. And the scheme that Uri Treisman developed for teaching his students to study has become legendary. He won a MacArthur Prize for his work. What is his methodology?

Treisman teaches his students to work in just the same way that professional mathematicians work. Namely, they work on their own for a while. Then they get together in small groups and exchange ideas and techniques. Then they go back to working on their own again. The cycle repeats until a particular homework assignment is completed.

A big part of mathematical maturity is *developing toward being a professional mathematician*. It follows that learning the study technique described in the last paragraph must be part of this process. If you cannot work efficiently and effectively and well, if you cannot study in such a manner that you are internalizing ideas and moving on to new ones, then you will not succeed in the scholarly game.[3]

---

[2] I teach at an elite private institution, and we typically attract top-ranking students with impressive SAT scores and other credentials. We have a program at Washington University called the Institute for Advanced Learning, and its purpose is to teach these highly qualified students how to study.

[3] I was once tutoring a student who was struggling with calculus. One day he came to me and said, "Each evening I sit in front of the book for three hours and nothing happens. I never get anywhere." I finally realized that that was a perfectly accurate description of what he was doing. *He was sitting in front of the book.* He was not thinking, he was not cogitating, he was not calculating. He was sitting and staring at the book.

Students need to learn that studying mathematics is not like studying kinesiology or philosophy. *You must study a mathematics book with a pencil in your hand*, and you must frequently turn aside and calculate and write and turn the ideas over in your mind. You must recast the ideas in your own language and figure out a way to internalize them and plant them in your brain. This is what Treisman taught his students, and a key part of his process was *teaching them to communicate with each other*. If you haven't thoroughly digested a mathematical idea, then you will not be able to communicate it. You can use communication as a tool to help you in the digestion process, and as you do so your words and thoughts will be hesitant and stumbling. Your goal is to be able to enunciate the ideas cogently and clearly and precisely to your peers. This activity goes hand in hand with being able to think straight.

Treisman's program at U. C. Berkeley was a great success, and his students went on to excel at calculus. He now runs an institute at the University of Texas where the Treisman teaching methods are developed and promulgated.

3.4   How Can We Improve the Education of Math Students?

There is considerable anecdotal evidence that the American school system does not encourage students to go into math and science. Some recent studies of education in different countries show that American high school students rank in the bottom quarter of participating countries, standing behind China, Finland, and Estonia.

A highly regarded assessment exam is the TIMSS (Trends in International Mathematics and Science Study). Developed by the International Association for the Evaluation of Educational Achievement (IEA), this study was first conducted in 1995 and has been repeated every four years thereafter. The most recent such study was in 2007. The TIMSS concentrates on 4th and 8th grades. In 2007, 48 countries participated. In the United States, 20,000 students were examined for mathematics skills. We ranked 11th of the 48 participating countries, after Slovenia and above Lithuania. The highest ranking countries were Singapore, Taiwan, South Korea, and Japan.

In another international study (involving 34 countries), the United States ranked 19th in science and 27th in math.

The development of our K–12 schooling is a complex issue, because there are many social forces at play. A college professor examining the situation is prone to concentrate on content. Are the students seeing the right stuff at the right time? Are there mechanisms in place to test student progress so that they are not allowed to proceed to the next level until they have mastered the current level? There are other powerful forces that shape our educational culture that are concerned with self esteem, with social progress, with empowering under-represented groups, and with encouraging women. Some responsible studies have advocated that, even if students do not master the ideas in grade X, they can still move on to the next grade because graphing calculators can cover for the deficiency (see [EC]).

The following parody of the evolution of thought on mathematics teaching has more truth to it than most of us would like to admit:

The Evolution of Teaching Math

Up to the 1960s: A peasant sells a bag of potatoes for $10. His costs amount to 4/5 of his selling price. What is his profit?

In the early 1970s: A farmer sells a bag of potatoes for $10. His costs amount to 4/5 of his selling price, i.e., $8. What is his profit?

1970s (new math): A farmer exchanges a set P of potatoes for a set M of money. The cardinality of the set M is equal to 10 and each element of M is worth $1. Draw 10 big dots representing the elements of M. The set of production costs C consists of 2 big dots fewer than the set M. Represent C as a subset of M and give the answer to the question: What is the cardinality of the set of profits?

1980s: A farmer sells a bag of potatoes for $10. His production costs are $8 and his profit is $2. Underline the word "potatoes" and discuss with your classmates.

1980s (alternative math): A kapitalist pigg undjustlee akwires $2 on a sak of patatos. Analiz this tekst and sertch for erors in speling, contens, grandmar and ponctuassion, and than ekspress your vioos regardeng this metid of geting ritch.

1990s: A farmer sells a bag of potatoes for $10.00. His production costs are 0.80 of his revenue. On your calculator graph revenue versus costs. Run the POTATO program on your computer to determine the profit. Discuss the result with the other students in your group. Write a brief essay that analyzes how this example relates to the real world of economics.

3.5   The Development of the Mathematics Curriculum

The mathematics curriculum at colleges and universities in the United States has not changed much in the past sixty years. The typical sequence of courses is

- Calculus
- Linear Algebra
- Differential Equations
- A Transitions Course (optional)
- Real Analysis
- Abstract Algebra
- Complex Analysis
- Differential Geometry
- Topology

Many students will not get through this entire list, but this is an ideal curriculum. [See also the Tree of Mathematical Maturity at the end of this book for a chart of mathematical development through coursework.]

There have been some tweaks to this model. Many students now get a fair dose of calculus in high school, so they can skip some or most of the first step. Many curricula now have added courses in Dynamical Systems and Fourier Analysis and Partial Differential Equations and Number Theory. Others have courses in Mathematical Modeling and Control Theory. As a result, the student no longer needs to follow the classical track of real analysis and abstract algebra. There are other choices.

But our basic view of what a mathematics education should consist of has remained fairly constant. The National Science Foundation (NSF) has had some (financial) initiatives to encourage math departments to completely re-think and overhaul the curriculum. I attended some presentations about this program, and what NSF had in mind was for people to set the usual program aside and to re-invent the math major from whole cloth. I do not know that any fundamental new ideas ever grew out of the NSF push.

It seems clear to me that, in the early 1960s (the days of the Sputnik era), universities were undergoing explosive growth. Many fast-growing departments came to the math department and told us that their students needed to learn some mathematics. What was to be done? We had neither the time nor the inclination to spend a lot of effort on these questions. And these other departments were ill-equipped to formulate or describe their needs. So we usually said, "Have them take calculus." The result is the massive calculus programs that practically every American math department has today.[4] At large state schools, the typical calculus enrollment in the fall semester is several thousand students distributed across ten or more lectures and hundreds of recitation sections. It is big business.

My own university has an accounting system (the *reserve system*, set up to please the medical school) which results in the phenomenon that we (the math department) are paid for each non-Arts-and-Sciences student that we teach. Put in other words: The math department is part of the College of Arts and Sciences. We are therefore obliged to teach calculus to Arts and Sciences students. But teaching students from the School of Engineering, or the School of Business, or the School of Architecture is not part of our formal purview. So we get paid for doing so.

Obviously this is a system that works to our benefit, and we are disinclined to want to change it. But, setting pecuniary temptations aside, it would make sense to re-examine the freshman and sophomore mathematics offerings from the point of view of non-math clients and see whether we can serve their needs more effectively. Would a business student be better educated by a course that taught a mix of statistics, combinatorics, and calculus? Would an engineering student benefit from a course that was part calculus and part differential equations? What about a physics student?

We could also re-examine how well the curriculum serves our mathematics majors. The old-school view of undergraduate education was that it was up to the students to battle their way through the curriculum, get their degree, and get into graduate school (or into a profession). Those who were not up to the task flunked out, and good riddance to them. This point of view is no longer considered acceptable. We want all of our students to successfully complete the undergraduate curriculum and go on to be happy and productive members of society. Today it is rather rare for a student to flunk out of college. There are too many mechanisms in place to provide support when the student has trouble, and to provide alternative paths when the student hits a roadblock.

---

[4]Conversely, calculus is the biggest thing in sight in the math department—from the point of view of departments that are fairly distant from mathematics (and also from the point of view of the administration). When another department thinks about math, it is natural for it to think about calculus. So when other departments came to us asking what their students should take, they may have already had calculus in mind.

If we had the faculty manpower to do so, we could have a special lower-division track (and upper-division too!) for future mathematicians. In such a track, we could begin early on to instill the values that lead to mathematical maturity. We might teach calculus from Spivak's classic book [SPI]. Then even students who have had calculus in high school will be seeing something new and challenging and exciting. We could teach a multivariable calculus course that integrates linear algebra in a serious fashion—so that the results (the chain rule, Green's theorem, etc.) can be proved in a mathematically natural way. And we could teach an ODE course that relies on linear algebra (so that the concept of general solution of a differential equation can be formulated properly) and basic real analysis. After laying such a foundation, we could teach advanced courses that were more honestly mathematical. The analysis course could have a serious treatment of differential forms. The algebra course could really nail the Jordan canonical form.

Unfortunately most math departments are stretched rather thin these days. We hardly have the manpower to cover the service courses that we are obliged to teach. As a result, the math majors are not getting the attention that they deserve. They *do* have to rely on their own resources to get what they really want and need from their education. We can only hope that in the future there will be more time to devote to these questions.

3.6   Learning on the Internet

Five or ten years ago, if one opened up an issue of *The Chronicle of Higher Education*, one encountered page after page after page of advertisements for OnLine universities and OnLine learning systems of one sort or another. Each ad was accompanied by a handsome portrait of some famous teacher who was eager (presumably for a tidy fee) to endorse this particular system.

Interestingly, these ads have almost completely disappeared. Why is that? It is because the big employers in this country—Microsoft, General Electric, General Motors, etc.—have let it be known that they want to hire young people who have been educated in the traditional fashion—*not* on the Internet.

It seems to be more evidently true in mathematics than in other disciplines that the path to success is to study at the feet of a master. Fields Medalist Steve Smale studied at Harvard with Raoul Bott, David Mumford studied with Oskar Zariski, Terry Tao studied at Princeton with Elias Stein, Jean-Pierre Serre studied in Paris with Henri Cartan. If Mumford had decided not to go to Harvard, but had instead studied at the University of

Southern North Dakota,[5] it is likely that his life would have followed a different path.

You cannot master a recondite field like mathematics with the laid-back sort of learning that comes with the Internet. What is lacking is the dynamic give-and-take between student and teacher—the asking of tough questions and the brave and energetic search for answers. It is possible that there are more workaday lines of endeavor—like selling real estate or landscaping gardens—where an Internet education would be more than adequate. But, when I am old and sick, I hope that the cardiologist who is giving me a quadruple bypass will not have been educated on the Internet. I would hope that the good doctor was trained at the side of a master in the field—one who can demonstrate all the moves first-hand, and where the mistakes lie as well.

The lesson for us here is that mathematical maturity is not fostered by reading pages on the Internet. Sure, students can garner some facts and perhaps even learn the statements and proofs of some theorems. But they cannot get the hands-on experience that is essential to growing your brain and honing your skills. They cannot get that sharp edge that really makes you a mathematician.

3.7 Capstone Experiences

In part because of pressures from the university accreditation agencies, math departments today are paying more attention to capstone experiences for their majors. What does this idea consist of?

By its very nature, the undergraduate math curriculum is disjointed. Upper-division students take courses in analysis and algebra and geometry and dynamical systems and number theory and there is no effort made—indeed no context for such an effort—to integrate the ideas together. How do these different concepts and techniques interact with and support each other? How does algebra come up in the context of analysis, or vice versa? The purpose of a capstone experience is to give students an opportunity to engage in a project that will help them to see some of these hidden connections.

Of course a capstone project will have a faculty advisor, and it is the job of the advisor to assess the student's abilities and to choose a topic that will be both satisfying and challenging. Sample capstone topics are

- The fast Fourier transform
- The Gauss-Bonnet theorem

---

[5] A suitable fictitious public institution of higher learning.

- de Rham cohomology

- The Stone-Weierstrass theorem

- The Baire Category theorem

- The classification of closed 2-manifolds

- Elementary group cohomology

- Hyperbolic geometry

You can see that each of these topics involves analysis and algebra, or analysis and geometry, or algebra and topology in interesting new ways. Each topic will force the student to read a variety of sources, and to synthesize them. The student will need to think about mathematics in new ways, and make new connections.

As a capstone mentor, you have the opportunity to help students see the curriculum as *not* broken up into disjoint pieces. You can teach students that mathematics is a broad, seamless cloth with many wefts and warps that interact in fruitful ways. You can show students how to draw on different parts of the subject and make a whole that is greater than the sum of its parts. You can make their education more than what it was by demonstrating what a rich and varied subject mathematics truly is.

Surely the capstone experience will make a positive contribution to the process of mathematical maturation. It will teach students critical thinking skills, and cause them to ask new questions. It will *not* necessarily teach students how to prove theorems, but there is much more to the mature mathematical life than that.

3.8 Challenging Your Students

A very easy trap to fall into these days—especially if you are caught up in teaching a lot of large calculus lectures or the like—is to make things as easy as possible for your students. Instead of setting a high bar and teaching students how to reach for that bar, you instead endeavor to spoonfeed everything to them. Just to make the path for everyone a bit easier.

If that is the case, then it is really too bad. The reason that you are a university professor, that you merit an exalted position in society, is that you hold the education of our youth in your hands. It is your job to figure out how to take these unmolded lumps of clay and turn them into dynamic, productive members of society. And part of that process is to teach them to bang their heads against the ideas, to mold and shape themselves by stretching their brains.

Certainly, if you want to be a part of the mathematical maturation process, then you need to get in there and get your hands dirty. Get inside your students' heads. Figure out where they are coming from and where they want to be going. Start by setting small hurdles for them, and work up to larger hurdles. Let them think about reaching for the stars by first reaching for the gleams of light that you set before them.

Helping students along to mathematical maturity is a rather personal process. You are going to have to get involved with them on a personal level. You want them to look up to you, and to aspire to be you. You want them to have a set of goals, and these should be goals that you create together and shoot for together. The students will depend on you, and you will be there as their inspiration and source of strength.

The way that you challenge your students should be a friendly process, but it should be a strict process. You want them to take the challenges seriously, and you want them to work hard to overcome them. You want them to understand that you are there to help at every step of the way. You are not only a source of information but you are a guiding light and a role model. You can show them what it is to be a mathematician.

3.9   The Sports Analogy

There are those among us with a surfeit of athletic talent. These individuals can run, jump, shoot, throw, slide, and break with incredible finesse and skill. But, in order to become effective winners, they must be coached. Pancho Segura played a key role in making Jimmy Connors a world-class tennis player. Bill Bowerman was Steve Prefontaine's mentor in running. Butch Harmon and Hank Haney helped Tiger Woods become a pre-eminent golfer. Dennis Ralston (along with Chris's father) were essential to Chris Evert becoming a frontline tennis star. Each of these coaches was a very special talent, equipped with drive, tenacity, technical skill, and an innate ability to teach these qualities. I should stress also here that each of these "learners" had a huge amount of natural talent; they just needed the proper mentor to help channel and focus that talent and turn it into an effective weapon.

Just as those who listen to popular music often do not appreciate the role of the arranger (Quincy Jones had a sterling effect on many modern hits; Art Garfunkel arranged most of the Simon and Garfunkel hits), so also many sports fans do not realize the essential role of coaching. The coach is the driving force, the source of skill and knowledge, and the inspiration behind the star.

And it is often just so in mathematics. There are certainly exceptions—
Ramanujan was largely self-taught, as were Abel and Galois. But it is really
no surprise that most of the Fields Medalists were students of absolutely
world-class mathematicians: Charles Fefferman and Terence Tao studied
with Elias M. Stein, David Mumford and Heisuke Hironaka studied with
Oskar Zariski, Jean-Pierre Serre and René Thom studied with Henri Cartan, Daniel Quillen and Stephen Smale studied with Raoul Bott, Laurent
Lafforgue and Ngô Bào Châu studied with Gérard Laumon, and on it goes.
It should be noted that Stein, Zariski, Cartan, Bott, and Laumon were *not*
winners of the Fields Medal. But they *taught* Fields Medalists (in fact they
each taught two, which is another record!).

It is also no surprise that most of the mathematical leaders of today studied at Princeton or Paris or Harvard or Berkeley or Chicago—because *that*
is where the fire is. That is where many of the key modern ideas are con-
ceived, and that is where their progenitors live and work.

You can learn mathematics and study mathematics and do mathematics
almost anywhere. And you can be happy doing so. But, just as it is valuable
to be able to locate the source of the Nile, so it is epistemologically useful
to know where mathematics originates, and how it flows.

3.10   Pacing

In Section 5.12 below we discuss what a good teacher does for students.
One of the points noted there is that the teacher sets a pace for the students.
This is an important concept, and worthy of further discussion.

When children are learning to play the piano, they naturally want to play
favorite tunes right away. But the correct way to do it is for the student to
first learn hand positions, then learn scales, then learn very simple tunes
played with one hand, then to begin to learn to play with two hands, and
so forth. It may take a couple of years before the student can begin to play
popular or familiar melodies. And a few more years before the student can
begin to show any real proficiency.

It is the same way with mathematics. Everyone would like to be Evariste
Galois and start proving earthshaking results when they are eighteen years
old. But it generally does not work that way. Mathematical maturity develops in many stages, over a protracted period of time. It does not work to try
to force young people to do research; they simply are not ready for it. They
are liable to find it disconcerting, and even terrifying.

I participated in a National Science Foundation Research Experience for
Undergraduates (REU) program when I was still a lower-division college

student. I realized after several weeks that I was in over my head. I could understand my research problem (at least I understood what the question asked), but I had no idea how to muster the initiative, focus the necessary technique, and develop the tenacity to make any headway on the problem. I found the entire experience to be frustrating and intimidating. I was depressed for the entire academic year following that experience, and came very close to quitting mathematics. I finally returned to mathematics because I found that it was the only subject that gave me the intellectual satisfaction that I craved.

By the end of my junior year I had taken some graduate courses and had been exposed in a natural fashion to some research problems. I had even solved a couple of them. By then I was perhaps ready for an REU experience. And I was admitted to an REU program for the following summer. I decided instead to spend the summer in Europe, and that was probably a good decision (for my mental health, if nothing else).

I am now (arguably) a mature mathematician, and am quite comfortable doing research. In fact that is my life. And I can even train others to do it. But it was a long road to get here, with many steps (and also mis-steps) along the way. That is the nature of the beast.

3.11   Problems for Students

Of course a big part of undergraduate education in mathematics is tackling problems. As freshmen we look at little problems. Routine problems. Exercises really. As sophomores we may graduate to idea problems, open-ended problems, conceptual problems. By the time we are juniors and seniors then, ideally, we should be thinking about theoretical problems. Moving further ahead, it can be said that professional mathematicians also think about problems. But they are thinking about much larger issues. In many cases, they are endeavoring to shape an entire theory, or view a subject from a new perspective. It may take years for professional mathematicians to achieve even some of their goals. And there are liable to be many missteps along the way.

A good teacher knows how to create problems for (undergraduate) students. The instructor will know how to craft a problem at just the right level, with just the right amount of challenge, and just the right amount of mystery, for each student. These problems will not generally be research problems. They will be known mathematics. But they will *feel* like research to the students. They will consume them in just the same way that our research consumes us. They will give students a taste of the mathematical life.

It is just a fact that students do not have the wherewithal to create problems for themselves. The student has neither the insight nor the experience nor the tools at hand to put together interesting or challenging questions. This assertion is true both of a freshman calculus student *and* of a Ph.D. student. There are precious few third-year graduate students who can competently and effectively choose their own thesis problems. After all, a good thesis problem has to be new, interesting, at just the right level, accessible, and it has to have the nature that, if things do not work out, then the thesis advisor can modify it a bit and turn it into something that will pay off. What is more, a *really good* problem should be something that will lead students to useful avenues *after* they have left school. Because we do not want any student's program to shrivel up and die right after the umbilical cord is cut. In short, there are many parameters to designing a good problem. How can a student possibly know how to do this? This is why we have thesis advisors.

Ultimately students, if indeed they are to become freestanding mathematicians, must learn how to create questions. And that is a major challenge. One of the reasons that the vast majority of math Ph.D.s in this country never publish a paper and never initiate a research career is that they cannot figure out how to establish their own intellectual independence and start their own research path. A big part of starting your own research career is learning how to find problems to work on. One way to do this is to attend a lot of conferences and talk to a lot of people. And go to a lot of seminars. Mathematicians are generous in sharing their ideas, and tossing out problems. You can pick up lots of good ones if you just listen and think.

But there is a real art to finding problems that are right for you. It is a little bit like being a popular recording artist. One of the keys to success for such a singer is finding just the right songs for your style and your image. The right backup band and the right arrangements are also important. But if you are country singer Shania Twain and you are trying to pump up your career with recordings of Schubert Lieder, then you are probably not going to get very far. Just so, if you are a new Ph.D. in several complex variables then you must find problems that **(a)** will fascinate you, **(b)** will allow you to make incremental progress and to publish partial results as you push towards your bigger goals, and **(c)** be of interest to others so that you can give talks and convince people that you are a productive, vital member of the community. None of us is born with an innate ability to navigate this tricky course. It is always helpful to have a mentor—even when you are a junior faculty member—to help you steer a sure path.

Part of mathematical maturity is learning this process of creating and/or identifying new problems to work on. If you are in several complex variables and decide to work on the corona problem, then you are almost certainly doomed to failure. For all the major people in the subject have put a good many years into this problem and gotten nowhere. This probably is not the right way to start your career. If you are in analytic number theory and decide to prove the Riemann hypothesis, then you are probably doomed to a similar ill fate.[6] In the case of the latter problem there are lots of little cottage industries that have sprung up so that you can at least make some contributions and show the world that you are alive. For the corona problem I would say the picture is less optimistic.

In a dream world we would all work on the hard, ideal problems that fascinate us. But there is a practical aspect of life as well. If you want to get a grant, if you want to be invited to conferences, if you want to get tenure, if you want to have the admiration and respect of your colleagues, then you must *produce*. You must write a reasonable number of papers and get them published. One aspect of mathematical maturity is understanding this rubric and carrying it out in your own life. Again, speaking strictly practically, the sensible thing to do is to save the big problems for your senior years in the profession. Once you are a tenured full Professor, then you have the latitude to work on a really hard problem without producing a lot of partial results along the way. That is what tenure is for. That is why the university system is set up the way that it is.[7]

So the point is that you need to plan your mathematical career. At each stage, you must choose problems that are just the right size and shape. At each juncture, you must be able to make a meaningful contribution so that the profession can see that you are a player. You must be able to write papers and publish them. It is *not* enough to say, "I have a twenty-step[8] program to prove the Riemann hypothesis. Check back with me in thirty years." Doing so, you will quickly find yourself marginalized, and you will no longer have a meaningful role in the profession. You must exercise some survival techniques if you are going to have a meaningful and fulfilling career.

---

[6]After proving the independence of the continuum hypothesis—and winning a Fields Medal for the result—Paul Cohen dedicated himself to proving the Riemann hypothesis. Even though he was evidently the smartest guy around, he failed. Cohen never even wrote a paper about the Riemann hypothesis.

[7]Of course there are notable exceptions to this advice: David Mumford, Barry Mazur, John Milnor, Terence Tao, Niels Henrik Abel, Evariste Galois, and a number of others attacked awesome problems when they were young (Galois, after all, only lived to be 20 years old and Abel only lived to be 26) and achieved impressive success. But it is probably unrealistic for most of us to model ourselves after these geniuses.

[8]One wag has asserted that such a program might consist of "Count to nineteen and then prove the Riemann hypothesis."

3.12   Student Research Problems

It is quite fashionable these days to promote undergraduate research. The laboratory sciences have involved undergraduate students in their research activities for many years now, and it is natural for them to do so. For a laboratory has a complex infrastructure that requires contributions at many different levels.

Mathematics is different. We think of ourselves as single-combat warriors. Our inclination is to sit in our offices and prove theorems. How can I get students involved in my studies of the automorphism groups of pseudoconvex domains? They do not even know the vocabulary.

But my saying this surely reflects my own personal shortcomings. Why cannot I, like a physicist, develop "toy" versions of my problems and teach them to students? Why cannot I train students to perform experiments that will inform my more sophisticated investigations?

One can argue that none of the living Fields Medalists did undergraduate research (there are a couple of exceptions, but let me ignore them for the moment). So who are the rest of us to be trying to plow such a path? Well, the living Fields Medalists were trained by traditional people according to a traditional paradigm. It probably never occurred to anyone to have Charlie Fefferman or Terence Tao working on a junior version of somebody else's research problem.

The American government is putting considerable resources behind undergraduate research. In addition to the venerable Research Experience for Undergraduates (REU) program of the National Science Foundation, there is funding through individual grants and through other programs to create summer internships and other encouragements for students. We would be wise to take advantage of these new resources and, in doing so, to help enrich our students' lives. We should not think that we are actually launching them on freestanding research careers. Rather, we are giving them exposure to the research life. We are giving them a close-up glimpse of what we do and how we do it. We are showing them mathematics outside the classroom—as a living, breathing intellectual entity.

I have run NSF-sponsored undergraduate research programs, and I consider them to have been a success. I tried to teach the students how to collaborate with each other, to communicate mathematics, and to engage in intellectual discovery. I taught them how to present their ideas at the blackboard. I tried to teach them to read the literature and to digest advanced ideas. I tried to show them how to think about a sophisticated problem and to chip away at it. Very few of my advisees in these programs actually solved their problems, and only a handful ever published anything. But a great many of them, based on their experience in our REU program, decided to go to graduate school in mathematics. Many of them earned Ph.D.s and are now productive faculty members at good schools. So by that measure we can consider our undergraduate research efforts to have been worthwhile.

Undergraduate research is certainly a situation in which we cannot expect the students to find their own problems. I know some graduate Ph.D. mentors who will tell a student, "Go read some of my papers and see whether that gives you any ideas for a research problem." That is OK I guess, but it tends to be rather hit-or-miss. My personal opinion is that the thesis advisor should take a more active role. But my main point here is that you simply cannot do this with an undergraduate. It is like telling a beginning piano student to go find a mazurka and then to figure out how to play it.

The first time I met my thesis advisor he gave me three research problems to think about. They all turned out to be way too hard for me (in fact they were all solved much later by teams of experts and all turned into major research programs that are still alive today). But this chicanery was his way of getting my feet wet, and showing me what research was like. The second collection of problems that he gave me were also too hard (although I ended up solving one of those some years later, and I turned that one into an entire research area). But, after some years and some struggle, we finally found a problem that was just perfect for me and that I could solve. This ordeal was a trial-by-fire approach to graduate education, but I got a lot out of it. By the time I was finished with graduate school, I was really ready to get out there and fight. I had both sunk and swum, and I had learned the difference between the two.

3.13 The Ph.D. Advisor

If there is anyone who is on the line in the mathematical maturity game it is the Ph.D. advisor. This individual—usually a professor at a university—is expected to take a bright student, who has spent his/her life studying *other people's* mathematics, and turn that student into a functioning mathematician who can create *his/her own* mathematics. There is hardly any transformation in life that is more dynamic or daring. It is almost like helping a snail to become a bird—transforming something that crawls along the earth into something that soars through the skies.

Along the way there will be many stumbles and missteps (both for the student *and* for the teacher). The student will try many things that will not work. It will be up to the professor to show the student how to learn from mistakes, to take something that does not seem to make sense and turn it into a useful tool. Students need to learn to read the literature and not only understand what authors are saying but also see how to pry out the key ideas and apply them to their own problems. Students must understand what authors are saying, but also understand what authors are *not* saying (but keeping to themselves).

A large part of the role of the Ph.D. advisor is to give students the right frame of mind. Teachers should imbue students with optimism and, perhaps more importantly, with confidence. They can do so by saying the right words, but also by breaking problems up into steps that students *can actually do*, so that they can *see* that progress is being made. This is a nontrivial but essential part of making the long trek towards the Ph.D. actually happen. It is a mechanism for effecting mathematical maturity.

Successful Ph.D. students, those who actually make it to the Ph.D., literally owe their (professional) life to their thesis advisor. They have a career because of the work done with this professor. It is a marvelous and life-changing experience.

CHAPTER 4

Social Issues

Did you mean to say that one man may acquire a thing easily, another with difficulty; a little learning will lead the one to discover a great deal; whereas the other, after much study and application, no sooner learns than he forgets?

Plato (philosopher)

Be sure to show your work for partial credit. You guys live and die by partial credit. In fact, everybody lives and dies by partial credit. Nobody gets anything right anymore.

Albert Stralka (mathematician)

One should always generalize. [Man muss immer generalisieren].

Carl Jacobi (mathematician)

Age is no guarantee of maturity.

Lawana Blackwell (author)

All newly discovered truth passes through three stages. First, it is ridiculed. Second, it is violently opposed. Third, it is accepted as being self-evident.

Arthur Schopenhauer (philosopher)

Mathematics is the science in which we never know what we are talking about or whether what we are saying is true.

Bertrand Russell (mathematician, philosopher)

4.1 Chapter Overview

As with any human endeavor, the study of mathematics is affected by human issues. We may wonder whether Asperger's syndrome, or schizophrenia, or manic depression are part and parcel of mathematical talent. Must

63

the developing math teacher become acquainted with these syndromes and learn how to deal with them?

We also may take an interest in various standardized means of measuring intelligence. The Stanford-Binet IQ test has been widely used to measure intelligence. The SAT and ACT Exams are still used to determine college placement. The Myers-Briggs Index is a standard device for classifying personalities.

What is the role of all these indicators for the math teacher and the developing math student? Should we be guided by such tests or should we chart our own path?

These are important—potentially life-changing—questions. We shall consider some of them in the present chapter.

4.2 Math Anxiety

About twenty-five years ago the phenomenon of "Math Anxiety" was identified and described—by well-meaning people, educators endeavoring to explain why some people have more trouble learning or dealing with math than others (see [TOB] and [KOW] for both history and concept). We do not hear much about math anxiety in math departments because such departments are full of people who do not have it. Math anxiety is an inability by an otherwise intelligent person to cope with quantification and, more generally, with mathematics. Frequently the outward symptoms of math anxiety are physiological rather than psychological. When confronted by a math problem, the sufferer has sweaty palms, is nauseous, has heart palpitations, and experiences paralysis of thought. Oft-cited examples of math anxiety are the successful business person who cannot calculate a tip, or the brilliant musician who cannot balance a checkbook. This quick description does not begin to describe the torment that those suffering from math anxiety actually experience.

What sets mathematics apart from many other activities in life is that it is unforgiving. Most people are not talented speakers or conversationalists, but comfort themselves with the notion that at least they can get their ideas across. Many people cannot spell, but rationalize that the reader can figure out what was meant (or else they rely on a spell-checker). But if you are doing a math problem and it is not right, then it is wrong. Period.

Learning elementary mathematics is about as difficult as learning to play *Malagueña* on the guitar. But there is terrific peer support for learning to play the guitar well. There is precious little such support (especially among college students) for learning mathematics. If the student also has a mathematics teacher who is a dreary old poop and if the textbook is unreadable, then a comfortable cop-out is for students to claim math anxiety. Their friends will not challenge the assertion. In fact they may be empathetic. Thus the term "math anxiety" is sometimes misused. It can be applied carelessly to people who do not have it.

The literature—in psychology and education journals—on math anxiety is copious. The more scholarly articles are careful to separate math anxiety from general anxiety and from "math avoidance." Some people who claim to have math anxiety have been treated successfully with a combination of relaxation techniques and remedial mathematics review.

It would be heartless to say to a manic depressive, "Just cheer up," or to say to a drug addict, "Just say no." Likewise, it is heartless to tell people who exhibit math anxiety that in fact they are just lazy bums. At the same time, mathematics instructors are not trained to treat math anxiety, any more than they are trained to treat nervous disorders or paranoia. If a student told you that he/she had dyslexia, then you would not try to treat it yourself; nor would you tell the student that he/she just didn't have the right attitude, and should work harder. Likewise, if one of your students complains of math anxiety, you should take the matter seriously and realize that you are not qualified to handle it. Refer that student to a professional. Most every campus has one.

Never forget that you are a powerful figure in your students' lives. This fact carries with it a great deal of responsibility. If you were a follower of Dr. Jack Kevorkian (1928–2011),[3] then you might take a troubled student in hand and say, "I know you are doing poorly in your math class. You must be in a great deal of pain, and suffering from shame. I have a solution for you—it's rather permanent, but it's painless." Chuckle if you will, but this is no joke. Problems of the psyche can be severe and dangerous. If any student comes to you with psychological problems then make sure that he/she gets help from someone who is qualified to administer that help.

Unfortunately, at some schools the "math anxiety" thing has gotten way out of hand. There are good universities where students may be excused from a mathematics or statistics course (one that is *required* for their major) by simply declaring themselves to be math phobic, or possessed of math anxiety. It is a sad state of affairs, but there is nothing that we math teachers can do about it. Because you and I are, by nature, good at mathematics,

---

[3]Dr. Jack Kevorkian was the physician who garnered considerable notoriety for assisting the suicides of severely ill people.

and because we do not suffer from math anxiety, it is difficult for us to empathize with people who suffer from this malady. It is best to let those who know the literature, know the symptoms, and know the treatments to handle students who have this form of stress. Do not hesitate to refer your students to the appropriate counselor when the situation so dictates.

4.3   Mathematics and Diversions from Mathematics

It would be wrong, and actually rather foolish, to suppose that successful mathematicians must have tunnel vision. That the only thing that they are interested in, or can be interested in, is mathematics. Most mathematicians are multi-dimensional creatures.

Many mathematicians are high-level musicians—actual performing artists. They take a good deal of satisfaction from the obvious connections of music with mathematics, but they also simply enjoy the music because it is beautiful.

Many mathematicians are artists—sculptors or painters or mosaicists. Many others are ardent rock climbers or scuba divers or triathlon runners. There are mathematicians who are gourmet chefs, and others who work with disadvantaged kids. There are mathematicians who are opera fanatics, and others who are animal rights activists. Many mathematicians make furniture or build houses or blow glass. Some are passionate about orienteering or skiing.

Some mathematicians spend their leisure hours studying and engaging in folk dancing. Others are semi-professional or even professional singers. A few are even rock musicians.

The lesson here is a simple but important one. Mathematicians are people. They have many aspectss, and many interests. They have a variety of activities to balance out their lives and keep themselves sane.

Many mathematicians also have a spouse or a life partner, and they have activities that they share with that partner. Such activities could include travel, or attending concerts, or going to gourmet restaurants, or taking arts and crafts classes together. This is how life ought to be lived, and it is as good and rewarding for the mathematician as it is for anyone else.

Not long ago I was visiting another university to give a talk. At the ceremonial dinner, one of the faculty wives apparently became annoyed at all the attention I was getting. She turned to me and asked, "So I guess you don't do anything but mathematics? You have no other interests." Naturally I wanted to smack her. But I instead gave some polite answer and then turned to talk to the person on the other side of me.

What I describe in this section is not mathematical maturity by any means. But it is a context for your life in which mathematical maturity can develop and flourish. It is a means to strike a balance, and keep yourself on an even keel.

4.4 National Standards

There are several organizations devoted to developing national standards for mathematics education. This is an interesting and potentially important development in our profession, and is worth pondering for a moment.

The tradition in the United States—until very recently—has been *local control of schools*. There were a variety of political and sociological reasons why this tradition took root in our country. Some of these had to do with creating institutional means for keeping certain minority groups in their place. Others had to do with religion, or with locally held social values. But one upshot of "local control of schools" was that public education in New York City was quite different from public education in Tuskegee, Alabama. And the residents of both communities would have said that that was the way they wanted things to be. Denizens of New York would perhaps have wanted their kids to become sophisticated and productive professionals in our society. (Caucasian) residents of Tuskegee would have said they wanted their kids to become genteel members of traditional Southern culture; African-American residents may have wanted something different. They wanted life to be *plangent*.

Matters have been different in Western Europe, for instance. In Italy, educational matters are decided in Rome. In France, education questions are dealt with in Paris. The good feature of such a centralized system is that there are national standards promulgated by educational professionals in the nation's capital. The bad feature is that there is little sensitivity to local needs, or to differences among different parts of the country. Just as an example, when the advanced degree of Ph.D. was finally introduced in Italy about twenty-five years ago, the government would each year send an edict to each Ph.D.-granting institution of higher learning that said, "OK, in this academic year you can grant five Ph.D.s." (Or substitute some other positive integer for five.) The point is that the government did this without any regard for how large the institution was, how many graduate students it had, or how many students were expected to complete the Ph.D. program that year.

The modern view in the United States is that we do not want young people growing up in rural Mississippi to be a priori shut out of going to a good college. We want them to have the option for a career in the high tech sector, or a life working on the genome project, or a path into the legal profession. We want everyone to have the same educational opportunities (whether they choose to take advantage of those opportunities is a different question). We want to have a mechanism to ensure that teachers in all parts of the country are well trained and well qualified, and will transmit their knowledge and expertise to their young charges. Thus we have the National Council of Teachers of Mathematics (NCTM) and the Common Core State Standards Initiative. These are two of the more prominent organizations that are directly involved with developing national standards in mathematics (and other disciplines as well) education. Part of what is interesting about these groups and their activities is that they are *not* sponsored by the federal government. The Common Core group is run, and funded by, the states. The NCTM is privately funded.

Going to the Web site

```
www.nctm.org/about/content.aspx?id=952
```

we find that NCTM has this among its goals:

> Additionally, it is suggested that future teachers need a "knowledge of the mathematics that students are likely to encounter when they leave high school for collegiate study, vocational training or employment" and "mathematical maturity and attitudes that will enable and encourage continued growth of knowledge in the subject and in teaching."

There is that buzz phrase again.

Going to the Web site

```
www.corestandards.org/the-standards/mathematics
```

we find that Common Core has this among its goals:

> One hallmark of mathematical understanding is the ability to justify, in a way appropriate to the student's mathematical maturity, why a particular mathematical statement is true or where a mathematical rule comes from. There is a world of difference between a student who can summon a mnemonic device to expand a product such as $(a + b)(x \mid y)$ and a student who can explain where the mnemonic comes from.

It is heartening to see these important groups discussing mathematical maturity with such confidence and such vision. But one of the reasons that the present book was written is that people use the phrase "mathematical maturity" without there ever having been a precise definition of what it is.[1]

4.5 The Myers-Briggs Index

The Myers-Briggs Index is commonly used to determine *what type of thinker* a person is. Is the person intuitive or analytical, sensitive or hard-nosed, conceptual or detail-oriented? The test proceeds by way of a set of seventy-two dialectical questions.

The Myers-Briggs Type Indicator was created by Katharine Cook Briggs and her daughter, Isabel Briggs Myers, during World War II. Based on the typological theories of Carl Gustav Jung, its goal was to help women entering the wartime industrial workforce to identify the sort of wartime jobs where they would be "most comfortable and effective." The original questionnaire of Myers and Briggs was used for just this purpose. It grew into the Myers-Briggs Type Indicator, which was first published in 1962.

Today mathematical advocates of Myers-Briggs assert that we have traditionally taught mathematics for just one personality type (analytical, judgmental, etc.) but that there are actually sixteen Myers-Briggs personality types (see the details below). Therefore we should learn to adapt our teaching to sixteen different ways of thinking.

Setting aside the practical difficulties with implementing such a program, we have to ask whether it is a valid point of view. After all, a good education entails exposure to different modes of discourse. Philosophy has one mode of discourse, biology another, and mathematics yet another. When we teach calculus, or set theory, or real analysis, we are (more than anything else) teaching students our mode of discourse. That entails axiomatics, logic, set theory, functions, and especially proofs. If a Myers-Briggs analysis tells us that our discourse is logical, cold, analytical, and so forth, then so be it. No doubt Myers-Briggs would say that the discourse of literary criticism is intuitive, social, sensitive, and so forth. And I doubt that the literary critics are about to throw up their arms and abandon what they do well. Why should we? The Myers-Briggs dichotomies are

Extraversion	(E)	(I)	Introversion
Sensing	(S)	(N)	Intuition
Thinking	(T)	(F)	Feeling
Judgment	(J)	(P)	Perception

---

[1]This is a little bit like fractal geometry—a popular movement in modern mathematics. There is no precise definition of "fractal," and that fact tends to detract from the precision and validity of the subject.

So we see that each person has one attribute from each row. Thus there are sixteen different types of personalities. Typical Myers-Briggs sentences (which are used in an aggregate assessment of seventy-two statements to do a Myers-Briggs analysis) are

1. You are almost never late for your appointments.

2. You are more interested in a general idea than in the details of its real-ization.

3. It's difficult to get you excited.

4. You often think about humankind and its destiny.

5. Objective criticism is always useful in any activity.

It seems that **1** is a J–P decider, **2** is an S–N decider, **3** is an E–I decider, **4** is a T–F decider, and **5** is a J–P decider. Myers and Briggs are not interested in how strongly the reaction to a statement moves you towards the lefthand choice or the righthand choice. For them it is a zero-one game.

The question for us here, in this book, is what does Myers-Briggs have to do with mathematical maturity? Are certain personality types more likely to become mathematically mature? And, if so, why?

I do not think that the answer is entirely obvious. I certainly know math-ematicians who are thinking and perceptive—sometimes even judgmental. Most people would say that a good many mathematicians are introverted. But certainly not all. There are some who are notably extroverted.[2] Quite a few mathematicians have wonderfully developed intuitions—especially (but not exclusively) in their areas of expertise. Many mathematicians are sensitive, and many are feeling. Many have well-developed personal lives.

So, on the one hand, the teaching-reformers who want us to teach mathe-matics in sixteen different manners because there are sixteen different per-sonality types are correct in noting that those different types really do come up in the context of mathematicians. On the other hand, that does not ar-gue that we need to teach mathematics in sixteen different ways. In point of fact, mathematics is a coherent subject that adheres strongly to a partic-ular paradigm and a particular mode of discourse. To abandon these would be to discard the core of what mathematics is (see [KRA2] for further dis-cussion of this point). It is a matter of whether the mountain will come to

---

[2] An old saw says that an introverted mathematician is one who looks at his shoes when he is talking to you. An extroverted mathematician is one who looks at *your* shoes when he is talking to you.

Mohammed or Mohammed will go to the mountain. Mathematics is the mountain. Students should go to it and learn its ways.

It is quite *au courant* to want to bend mathematics to help achieve certain social goals. These sometimes include:

- Increase students' self-esteem.

- Empower or enable certain under-represented groups in society.

- Increase the representation of women in mathematics.

- Ensure that students advance without fail through the learning cycle (sometimes regardless of achievement along the way).

- Increase the number of students who graduate.

- Increase the number of students who join the workforce.

- Decrease the number of student dropouts.

- Diminish the attrition rate.

All of these goals have merit. But they should not be conflated with the actual *teaching* and *learning* of mathematics. Those are intellectual endeavors which, in the best of all possible worlds, should be free of social pressures.

Perhaps it may be said that entries T, I, J, and P in the Myers-Briggs chart are more redolent of mathematical maturity than some of the other human attributes that are listed there. In this book our focus is not so much on characteristics of personality but rather on the nature of learning.

4.6   Intelligence Tests

The idea of testing for intelligence seems to have originated in China in the seventh century under the Sui Dynasty. The motivation was the identification of future civil servants and civic leaders. The idea of the Chinese imperial examination matured during the Tang dynasty, and was in place for nearly six centuries. It was ultimately abolished in 1905 by the Qing dynasty. The Chinese now have a new method for selecting civil service staff.

In modern times, the Englishman Francis Galton (half-cousin to Charles Darwin) developed a means using nonverbal sensory-motor tests to measure intelligence. While it initially found favor, it was ultimately abandoned because it seemed to have no correlation with expected outcomes such as performance in college.

In the late nineteenth century, the French psychologist Alfred Binet, working with Victor Henri and Théodore Simon, developed what came to be known as the Binet-Simon test. Their primary concern was the identification of children who were mentally retarded. American psychologist Henry H. Goddard translated the Binet-Simon test in 1910. Psychologist Lewis Terman at Stanford University revised the Binet-Simon intelligence scale and created what came to be known as the Stanford-Binet intelligence scale. The resulting Stanford-Binet test became the standard for intelligence testing in the United States for many years. Other well-known modern intelligence tests include the Wechsler Adult Intelligence Scale, Wechsler Intelligence Scale for Children, Woodcock-Johnson Tests of Cognitive Abilities, Kaufman Assessment Battery for Children, and Raven's Progressive Matrices.

Tests of this kind have come to be known as "IQ Tests," the acronym coming from the German phrase *Intelligenzquotient*. In the United States it was widely held (by educators as well as others) for many years that the IQ score of a student was predictive of performance in school, and indicative of aptitude for college. Not to mention predictive of success in adult life.

Today people have backed away from these views. It is now understood that a child's IQ score can depend on morbidity (proclivity towards disease), parents' social status, environmental factors, parental IQ, nutrition, and many other non-scholarly factors. It is easy to see that a test like this can be culturally biased: a student educated in an inner-city school with substandard textbooks and teachers, and with no exposure to cultural stimuli, would not even understand many of the questions. Statistics indicate that boys tend to do better on these tests than girls.

What do such tests have to do with mathematical maturity? IQ tests use a variety of different examination techniques. Some questions are visual, some are verbal, some questions only use abstract-reasoning, and some questions concentrate on arithmetic, spatial imagery, reading, vocabulary, memory, or general knowledge.

It is clear that some of these testing devices would detect a *potential* for mathematical growth. A strong performance on abstract reasoning and spatial imagery problems would indicate that the student has a bent for mathematical reasoning. But, for mathematical maturity, the proof is in the pudding. You are mathematically mature if you can do mathematics: solve problems, prove theorems, perhaps write and publish papers. No one would believe that you could detect a future concert pianist, or a future pioneering neurosurgeon, or a great poet, or perhaps a great inventor by looking at Stanford-Binet test scores. Just so with mathematics. Mathematical maturity entails too many intangibles. It is both a state of mind and a state of being.

4.7   Asperger's Syndrome

Asperger's syndrome is a form of adult autism.[3] Asperger's syndrome or Asperger's disorder is an autism spectrum disorder that is characterized by significant difficulties in social interaction, along with restricted and repetitive patterns of behavior and interests. It differs from other autism spectrum disorders by its relative preservation of linguistic and cognitive development. Although not required for diagnosis, physical clumsiness and atypical use of language are frequently reported.

Asperger's syndrome is named after the Austrian pediatrician Hans Asperger who, in 1944, studied and described children in his practice who lacked nonverbal communication skills, demonstrated limited empathy with their peers, and were physically clumsy. Fifty years later, it was standardized as a diagnosis, but many questions remain about aspects of the disorder.

Although individuals with Asperger's syndrome acquire language skills without significant general delay and their speech typically lacks significant abnormalities, language acquisition and use is often atypical. Abnormalities include verbosity, abrupt transitions, literal interpretations and miscomprehension of nuance, use of metaphor meaningful only to the speaker, auditory perception deficits, unusually pedantic, formal or idiosyncratic speech, and oddities in loudness, pitch, intonation, prosody, and rhythm of speech.

There is evidence that Asperger's syndrome is correlated with a highly developed and rarified intelligence. Some scholars believe that Glen Gould, Béla Bartók, Thomas Jefferson, Emily Dickinson, James Joyce, Adolph Hitler, Michelangelo, George Orwell, Wolfgang Amadeus Mozart, Alan Turing, Ludwig Wittgenstein, Andy Warhol, Jonathan Swift, Albert Einstein, and Isaac Newton had some form of Asperger's. Others would assert that there is not enough evidence to be able to say with any degree of certainty. Certainly Newton and Einstein had trouble with social interactions. Newton had almost no friends. Hitler had many pathologies. Mozart had extremely underdeveloped social skills, a short attention span, and limited verbal acuity. But that is not the same as having a clinical condition.

There is at least one living Fields Medalist who is known to have Asperger's. He does exhibit some of the strange behavior that is associated with the disease.

It would be incorrect to suppose that mathematical maturity has any correlation with autism or with Asperger's syndrome. Although there is some evidence that people afflicted with Asperger's can develop into powerful in-

---

[3] Although children, too, can suffer from Asperger's—see [HAD].

tellects, the converse is not true. It is the case that an all-consuming passion for mathematics can cause one to neglect other aspects of life—including personal grooming and consideration of others. Newton and Norbert Wiener used to forget whether they had eaten. Kurt Gödel starved himself to death. Thomas Simpson died of drink and depravity. Tycho Brahe died from over-consumption of beer. Victor Hugo used to do his work in the nude (but that was part of his regimen so that he would not be distracted). Similar negligence can be attributed to some musicians and some artists. But that is not the same as a clinical condition.

Some mathematicians are quite arrogant, and as a result have difficulty interacting with other people. Others are pathologically shy. Near the end of his life, Serge Lang told me that he had no friends left. [Of course he had had fights with all of them.] The explanation was really rather simple. This does not mean that he had Asperger's.

It would be a sad day if we were to conclude that those with mathematical ability, or those who possess mathematical maturity, are simply individuals who have a particular disease. If we look at the vast spectrum of people who have been successful mathematicians—from Archimides to Gauss to Cauchy to Weierstrass to Gödel—we can see many different personality types and many different intelligence types. Also many different styles of study and work. Mathematics is not a syndrome. It is an avocation.

4.8   The ABD

It is unfortunately all too common for a student in a math Ph.D. program in this country to decide to bail out. This could mean that the student will earn a Masters degree and then leave the program. Although less common in mathematics than in some other disciplines, it also possible for the student to earn an ABD ("all but dissertation") status—which is a formal designation of the fact that the student has completed all aspects of the doctoral program *except* for having written a thesis—and then to leave the program. Finally, there are some students who either cannot pass the qualifying exams or who become so disheartened that they simply quit, and leave the program with nothing to show for their efforts.

This sad state of affairs does say something about mathematical maturity. For it shows that said student quite simply cannot make the transition from learning mathematics that others have produced to *creating new mathematics*. Put in other words, the student cannot segue from being a mathematical dilletante to being a mathematician.

These are hard words, but what is described in the last paragraph is what the final, and most significant, stage of mathematical maturity is all about. Can you step into the ring and do what Gauss and Riemann and Cauchy and also what your professors have done—which is to produce new mathematics? Or will you spend your life just being an observer? It takes genuine courage and determination, and also some luck, to prove your first theorem, and then to find the means to prove more theorems after that. But that is the mathematical life. If you choose it, then you have to fight the good fight and make it happen.

4.9   Men and Women

Objective studies have suggested that women (in contrast with men) have, on average, greater language skills, better visual memory, clearer emotional judgment, and superior skill with mathematical calculation. Men, by contrast, are better at mathematical problem-solving, abstract reasoning, and visual-spatial awaremess.

One could argue at length about why this difference—if indeed it is valid—exists. There surely could be genetic, environmental, social, and other factors that play a role in women and men measuring differently for these factors.

But, taken at face value, it sounds as though men are more predisposed to mathematical maturity than women. *If* that is true—and I consider this to be a big "if"—then it could explain why there are more, and more successful, men in mathematics than women. It could explain why no woman has ever won the Fields Medal, none has won the Abel Prize, and none has won the Wolf Prize. Also relatively few have won the Steele Prize. It may be noted that quite a few women have been elected to the National Academy of Sciences in mathematics, and several have received the National Medal of Science. These are awesome encomia.

We all value our female colleagues, and appreciate their many contributions to the field. We would like them to be treated as equals in every respect. If there are biological differences between the genders, we need to understand those and learn how to come to terms with them. This is only in all of our best interests.

5

Cognitive Issues

Mathematical maturity is when you're grown up enough to handle a 2ϵ.

Michael Sharpe (mathematician)

A method in mathematics is a trick that is used more than once.

Ron Getoor (mathematician)

Everything is trivial when you know the proof.

David V. Widder (mathematician)

Obvious is in the eye of the beholder.

Ron Getoor (mathematician)

If we accept that mathematics is a sort of language, then I suppose that developing mathematical maturity is akin to becoming fluent in the language.

Harold Boas (mathematician)

When we were children, we used to think that when we were grown-up we would no longer be vulnerable. But to grow up is to accept vulnerability... To be alive is to be vulnerable.

Madeleine L'Engle (author)

5.0 Chapter Overview

"Nature versus nurture" is a very old question in the study of child development. Do children become scholastically gifted because of good genes, or because of a supportive environment at home (and at school), or both? Since the math teacher is more a part of the latter aspect than the former, it is natural for teachers to want to understand this dialectic.

There are many different types of learning: rote learning, learning by trial and error, learning by attacking difficult problems, learning by imitating a master, learning by exploration, learning by experimentation. Which of these is most relevant to developing mathematical maturity? Which is inimical to mathematical maturity?

If we are to be effective mathematics teachers, we should endeavor to understand students' values and students' goals. Not to mention their motivations.

All these ideas are explored in the present chapter.

5.1   Nature vs. Nurture

Studies of education and human intelligence (by Jean Piaget, Alfred Binet and others) indicate that many people are born with considerable intrinsic intelligence. But the environment in which a child is raised can also exercise a decisive influence over mental development. If the parents run a household that is linguistically sophisticated and stimulating, full of books, and rife with intellectual discourse, then the child will experience rapid verbal and cognate mental development. Otherwise possibly not.

Is there an analogue of this for mathematics? Many parents are math phobic, or at least have a rather negative memory of their mathematics experience in school. Many school math teachers are not comfortable with mathematics, and can do little more than read the math text to the students. This situation does not provide a nurturing environment for rapid or effective mathematical maturation.

Studies (usually performed by separating identical twins) indicate that nature plays a stronger role in the development and maturation of intelligence than nurture. See [HARR] for a careful consideration of the two points of view. But both aspects of development are important, and neither should be neglected.[1] A ghetto child who grows up in a household where both parents have two or three menial jobs gets little attention and virtually no nurture. Such a child often shows up at school not knowing how to hold a pencil, not knowing how to interact with other children, and even not knowing how to form sentences. Certainly a youth of this type is way behind the curve in mental development. Asking about mathematical maturity for such a youngster is completely out of the question.

5.2   Maturity vs. Immaturity

A mathematically mature individual internalizes big ideas, and these big ideas are an umbrella for many small artifacts, calculations, and methods.

---

[1] In fact there is considerable debate, much of it emotional, about this dichotomy. It is safest to listen carefully to both points of view.

A mathematically immature person merely knows some devices and algo-rithms.

This analysis is a bit simplistic. There are accomplished mathematicians who are very knowledgeable and know lots of abstract mathematical theo-ries, but they also know a lot of tricks that are useful in doing calculations and putting together proofs. And there are people with poorly developed mathematical skills who are still capable of coming up with useful ideas and insights.

But the gist of the first paragraph is certainly correct. Mathematical ma-turity is indicative of an ability to see the big picture, to use abstraction to group ideas together in useful ways, to be able to pass back and forth between unifying concepts and specific instances.

A big part of mathematical maturity is learning to pass ideas from your conscious mind to your unconscious mind. When you first encounter a new idea—say (to be a bit simple-minded) that

$$a^2 - b^2 = (a - b) \cdot (a + b)$$

—you must think about it and understand why it is true. Get clear why it works. After a while, you sublimate the concept. So that, the next time you see $a^2 - b^2$, you know instantly that it factors as $(a - b) \cdot (a + b)$. You no longer have to think about it.

This is a little bit like driving a car. If you had to think consciously about the steps to make a left turn (first hit the turn signal, then begin to turn the wheel to the left, reduce speed by lifting your foot from the gas pedal, turn the steering wheel sharply, straighten the car out, etc.), then you would not be a very effective driver. Nor a safe driver. Just so, if you have to think out every mathematical step or calculation that comes along, then you will never be able to point your brain at the big picture. Certainly you will never be able to think about high-level ideas like proofs. You could not consider doing research.

Thus we see, once again, that there is a layering of mathematics by level of difficulty or sophistication. In order to master level N, you must have internalized level $N - 1$.

5.3 Rote Learning vs. Learning for Understanding

As mathematics teachers, we are aware that there are different types of learning. The most mundane sort, the type that a student engages in just to get by, is *rote learning*. That model entails no understanding, no inspira-tion, no quest for knowledge. It is just a matter of hanging certain facts or techniques or tricks on hooks in one's head—and not for permanent archiving, but rather just as a temporary device to get one through an exam or other upcoming task.

The more thoughtful, engaging, dare I say *admirable* type of learning, the sort of learning that travels well and evolves into more learning, is learning for understanding. When one learns for understanding, one focuses on big ideas, and on grappling with the big picture. One sees the facts and techniques and tricks as specialized corollaries that follow naturally once one has the understanding in place.

Naturally part of the process of achieving mathematical maturity is leaving the first type of learning behind and moving firmly into the second type. Professional scholars concentrate on learning for understanding. Research scientists concentrate on learning for understanding. Literary critics specialize in learning for understanding.

In seeking to define mathematical maturity, one can say that it encapsulates a *style of learning and of studying*. It describes a person whose *métier* is mastery of big ideas and learning how to use them.

Although garnering knowledge comes with the territory, knowledge in and of itself is *not* mathematical maturity. If you do not know anything at all, then you cannot do much with the big ideas that you are learning. Certainly knowledge is essential to everything we do. It is our bricks and mortar.

But you develop your mind by beating it against things—namely, against big, hard ideas. Accumulating knowledge—facts and calculations—is a relatively passive activity and does *not* develop the mind.

There are certain cultures in the world today where rote learning is the tradition. And the style of teaching is adapted to that set of goals. The teacher stands at the podium and read the text to the class. *Literally.* The teacher does not tolerate questions. Students are taught to regurgitate knowledge, and little more. Although these countries have occasionally produced outstanding mathematicians, this is the exception rather than the rule. As these countries are becoming Westernized, things are changing. But it is a slow process, and there is considerable resistance.

5.4 How Do Students Learn? How Do Students Think?

Generally speaking, students do not have the big picture. They sign up for a class because they have heard that the professor is good, or they have heard that the subject is interesting. Perhaps even exciting. But they have no idea of how this course fits into the whole curriculum, or into their course of study. They do not know what the purpose of the course is, where it leads, nor what it implies.

This is part of your job as instructor—to help students to see why this course exists and what function it serves. If you are lucky, they will listen to you and come away with something more than what they had to begin with.

When working on the course—listening to lectures and (we hope) reading the book and doing the homework sets—the students proceed step-by-step. You cannot hope that they are reading ahead in order to get a larger idea of how things fit together. They learn the ideas one-at-a-time. In sequence.

And they tend not to ask hard questions. Their primary goal is to understand the examples so as to be able thereby to do the homework problems. In a more advanced course the connection between examples and homework is much more tenuous, but the general point of view is still at play. Undergraduate students do not have highly developed critical thinking skills. They are not taking the subject apart and putting it back together again. They are just going along to get along.

As I have said in many other contexts, one of your most important activities for your students is to serve as a role model. You show the students what it means to be an educated person, you show them how a mathematician behaves, and most importantly how a mathematician thinks.[2] In the best of all possible worlds, the students become educated by imitating you. As they begin to create their own proofs, they will do so by mimicking the proofs that you present on the blackboard. As they learn to formulate and establish counterexamples, they will do so by following closely the way that you present counterexamples in class.

Your hope as a teacher is to aid in developing the way that your students think. They will begin in the plodding fashion that we have described here, and you hope that, after spending 12 or 15 weeks in your august presence, they will have developed some acumen for mathematical thought. They will start formulating their own questions (not just questions from the textbook) and creating their own answers (not just answers from the back of the book). They will turn the ideas over in their minds and learn the means for internalizing them. They will see how to make the ideas their own.

---

[2]When I was an undergraduate I had at least one professor who never prepared his lectures—even in a very advanced course. He would come to class and screw around and make a *lot* of mistakes. At the end, by way of apology, he would say, "Well, now you have a lesson in how a mathematician thinks." This is really pretty silly. What we actually had was a lesson in how to be irresponsible and not do your job. There are more intelligent ways of showing people how a mathematician thinks.

I have frequently said that the only things in mathematics that I really know are things that I have discovered myself and taught myself. This may sound rather solipsistic. I certainly do not mean to take anything away from the many wonderful teachers that I have had. But I realized, after I became a faculty member at a good university, that nothing that I had seen in any of my classes—*especially* the undergraduate classes—had stuck to my ribs. It was like I took a bath in it, but then it all went down the drain. Now that I had to teach the stuff, now that I was responsible for it, my entire point of view changed. And for the better. I realized that I had to be able to understand things well enough to explain them cogently to struggling students, students who did not even have the argot to engage in mathematical discourse. And this need required rethinking everything from the ground up. It was a wonderful intellectual exercise, and it taught me so much.

One of the reasons that I so enjoy writing books is that it is a way of formalizing the process just described. It is one thing to re-invent and internalize a subject for your own satisfaction (and, we hope, for the benefit of your students). It is quite another to do so as part of a process of recording the subject in your *lingua franca* for future generations. Writing your material in a book truly makes it yours, and puts your stamp on it. Your name becomes associated with the subject—and for good reason. Now you are one of the established experts.

5.5 How Do Students Become Motivated?

We would all like to think that our students are miniature versions of ourselves—future mathematicians just clamoring to develop and mature. They are all sitting there in your calculus class just chomping at the bit, hungry for knowledge, ready to sink their teeth into the subject. Well, it is not true. Sorry.

Many students—in your calculus class particularly—do not even want to be there. They are taking the class because it is required for their major. Worse, the mentors in the major subject area never really make it clear to the students *why* they are being made to take calculus. As an extreme example, take pre-med students. Why must they take calculus? What does solving maximum-minimum problems have to do with treating patients and stemming disease? The answer is, "not much." Calculus is required of pre-med students simply to serve as a filter; the idea is that a student who cannot learn calculus cannot be very smart and therefore probably will not make a very good physician.

Even the students who think they may want to be math majors are not as well motivated as we would like. At the age of 18, their ideas about career paths—even education paths (Bachelor's degree followed by Masters degree followed by Doctorate followed by Postdoc followed by Assistant Professorship and so forth)—are vague at best. They are certainly aware that calculus is a key step in the progression of learning mathematics, but they have no sense of how the subject develops or how its different parts fit together. They want to do well in your calculus class, but they do not have the fire-in-the-guts that comes from firsthand knowledge of exactly what it is they are trying to achieve. Many of them are hoping—eventually—to be enlightened. But they are not enlightened yet.

Students become motivated by being enabled. They need to have their self-esteem pumped up, and this will in turn give them the confidence to want to forge ahead. I teach at an elite private institution that has truly gifted students. These kids are a pleasure to teach, and I enjoy not only teaching them but advising them too. So I am always a little astonished when bright graduating seniors tell me that their plans after graduation are to live with their parents for a couple of years in order to have some time to ponder the meaning of life.

What does this mean? Do you not want to jump-start your life? What do your parents think about this? Do they not want you to get out there and do something with this $200,000 education that you just completed? Where is your drive? Where is your tenacity? Where is your motivation?

Well, it is not there. Many bright students are quite timid about actually getting involved in life. Many are unsure of the challenges and disappointments that they may have to face. Without being able to articulate the thought, they know intuitively that school has been a rather artificial environment in which someone was always looking over their shoulder. The real world is a somewhat harsher place. If you fall down then you have to pick yourself up. Nobody is going to do it for you.

As Elizabeth Taylor said in the movie *Giant*, "I can raise my kids but I can't live their lives for them." We cannot get inside our students' heads and force them to grow up. What we *can* do is to give them the intellectual tools so that they can do it for themselves. Mathematical maturity is definitely *not* the same thing as emotional maturity or social maturity. But mathematical maturity can give you the confidence to make something of yourself. It can give you the insight to be able to chart a life path and follow it. It can be your source of strength.

5.6   Learning to Recover from Mistakes

There are few professions in which one makes one's living by learning from,
indeed recovering from, mistakes. You would never hear a physician say
(even though it may be true) that she makes a point of trying to learn from
her mistakes. A psychiatrist does not usually say that he tries to learn from
his mistakes. A civil engineer—one who builds bridges let us say—is not
usually heard at cocktail parties saying that she endeavors to learn from her
mistakes.[3]

Not so for mathematicians. Mathematician very rarely make a beeline for
the results that they seek. Instead, a mathematician tries things, formulates
guesses and conjectures, experiments, talks to people, realizes that his/her
point of view is incorrect (and adjusts it), and endeavors to push ahead.
Frequently the upshot of today's work is merely to realize that yesterday's
work was incorrect. Sometimes there will be a week of heartening progress
followed by a day in which it is determined that all the results are incorrect.
Other times there will be major breakthroughs that hold fast, followed by a
month of no progress at all.

This is the life of a mathematician. We definitely make mistakes on a
regular basis, and we train ourselves to learn from them. And a distinct,
identifiable part of mathematical maturity is understanding and mastering
this process. If you do not learn to make mistakes and then pick yourself
up and recover from them, then you will never be a mathematician. Mathematicians are *not* people who murder every problem that they encounter.
Instead, mathematicians are comfortable with the paradigm, "Two steps forward, one step back."

Henri Poincaré thought he had a proof that any closed, homotopically
trivial 3-manifold was homeomorphic to the 3-sphere. His proof turned out
to be incorrect, and this discovery led to the formulation of the Poincaré
conjecture—one of the biggest ideas of twentieth-century topology. This
fumble in no way diminished Poincaré's reputation. It was just part of his
discovery process. Augustin-Louis Cauchy thought that the pointwise limit
of continuous functions was continuous. He later realized his error and
helped to develop the theory of uniformly convergent sequences of func-
tions. Gottfried Wilhelm von Leibniz originally thought that $(f \cdot g)' = (f') \cdot (g')$. He made a similar error with the quotient rule. But he ultimately

---

[3]To be fair, there are some remarkable physicians who have written about the process of
learning from mistakes. These include William Nolen [NOL], Atul Gawande [GAW], and
Abraham Verghese [VER]. There are also engineers who have advocated this approach to their
subject—notably Henry Petroski [PET].

realized his mistake and learned from it. In the end he promulgated the *correct* product and quotient rules, and taught us a lot about calculus.

Serge Lang once told one of his Ph.D. students that he would never be a great mathematician because he was too afraid of making mistakes. It is certainly true that, if you want to make dramatic discoveries, if you want to "bend" your subject, if you really want to have an effect on your field, then you must be daring. You must be willing to try things. Doing so, you will certainly make mistakes. You probably do not want to advertise them (although, unfortunately, this can happen as well). But you *do* want to learn from them.

The Italian Jesuit Giovanni Saccheri (1667–1733) has been criticized for rejecting hyperbolic geometry because he found it "repugnant." But his tract *Euclides ab Omni Naevo Vindicatus* in fact opened the eyes of Lambert and possibly Gauss to the value of non-Euclidean geometry. Saccheri's book was not published until two months before he died, because he had to await approval of the Inquisition before he was allowed to publish. We note that Saccheri demonstrated here a real ability to learn from his errors. That is the sign of a true scholar.

Most of our errors do not get publicized. They do not go down in the history books. People do not prattle about them at weekend barbecues. They are just part of our own personal *angst*. This yoga is an integral part of how we do business. It is our learning process. Civil engineers and neurosurgeons must build failsafe mechanisms into their procedures because they cannot afford to have bridges that collapse and cannot afford to lose patients on the operating table. We mathematicians have a bit more latitude in how we proceed. Like chess grandmasters, mathematicians can offer to sacrifice their queens in pursuit of the entire game.[4] When we are lucky, the stratagem pays off.

5.7 Will These Ideas Travel Well?

Those who have mastered big ideas will have, as part of the process, determined how to communicate those ideas. After all, they have figured out how to communicate the ideas to themselves. The next natural step is to communicate them to others. This insight is why many of us decide to become teachers.

And big, all-encompassing ideas travel well. That is because they are flexible. They adapt themselves to many different languages and many dif-

---

[4]Here we are paraphrasing G. H. Hardy [HAR, Section 12].

ferent contexts. They hold substantial meaning for many different types of people.

Albert Einstein's ideas about special and general relativity have traveled to the far corners of the earth. They are everywhere, and they have had an immense impact in our lives. Just as an instance, Global Positioning Systems (GPS) could not function if we did not understand general relativity. For, at the altitude where the GPS satellite lives, gravity is weaker (than at the earth's surface) and therefore time moves at a different rate and compensation is needed—see [ASH].[5] There are many other examples.

Isaac Newton's ideas about calculus have permeated all aspects of scientific life, and they impact even those who know little or no mathematics. For the coding theory behind a CD disc or a DVD disc is all based on Fourier analysis, and that in turn is based on calculus.[6] Any part of modern technology that involves motion—especially acceleration—is based on calculus. Any rate of change is formulated analytically in the language of calculus.

By contrast, a trick for calculating cube roots, or a device for converting from arithmetic base 3 to arithmetic base 7, is of little interest and little utility. These are charming amusements, and may provide some mental gymnastics for a developing mind, but they have no real importance. One would not by nature give any real thought to how to communicate these tricks, or whom to communicate them to.

Mathematical maturity consists in part of learning the difference between big and small ideas. Burgeoning young mathematicians aim to leave the second type of ideas behind and to launch themselves into the challenging intellectual world of the first type of ideas.

5.8   Mathematical Maturity vs. Mathematical Inquisitiveness

It is difficult to do mathematical research unless you know something about mathematics. One cannot do serious investigation in any intellectual arena unless one has the appropriate tools at hand. For mathematics those tools include logic and smarts, but they also include the basic tenets of mathematics. If you want to explore geometry, you had better know some basic geometry. If you want to explore real analysis, you had better know some-

---

[5]This is really quite a spectacular instance of technological progress. It took decades after Einstein's original papers for there to be adequate experimental evidence for general relativity. Now the phenomenon is part of practical technology.

[6]Interestingly, the coding theory in a cell phone is based on the Cayley numbers—an entirely separate set of modern mathematical ideas.

thing about the real line, about compact sets, and about convergence of sequences and series. If you want to investigate partial differential equations, then at the very least you need to know some techniques for solving differential equations, and you also need to know the basic rubric of existence and regularity.

Put in other words, it is difficult to be mathematically inquisitive—at least in an effective way—unless you are mathematically informed. And it is just the same way with mathematical maturity. Mathematical maturity is not an abstraction that one can develop in the absence of traditional mathematical training.[7] You cannot take bright youngsters who have never had a math course and begin to develop their mathematical maturity.

This insight is why mathematics education proceeds in segments. A five-year-old child begins by learning arithmetic: addition and subtraction of whole numbers. Later comes multiplication. These are relatively straight-forward, virtually mechanical skills that entail few genuine ideas. Then comes division, and that is definitely interesting because it leads to fractions. It is generally agreed that one of the real hangups in K–6 math teaching is how to teach division of fractions.

We "mature" mathematicians know of course that "fractions" (or the *rational numbers* $\mathbb{Q}$) are in fact equivalence classes of ordered pairs of integers. Once one understands that construction, then *all* the arithmetic operations on $\mathbb{Q}$ can be resolved by fiat; you simply *define* addition, subtraction, multiplication, and division of rational numbers in the language of the equivalence classes. And all the desired properties of these arithmetic operations can be verified by straightforward proof.

The point is that the built-in ambiguity that arises from the fact that

$$\frac{1}{2} \quad \text{and} \quad \frac{3}{6} \quad \text{and} \quad \frac{8}{16}$$

are the same number is swept under the rug by the equivalence class construction. We can verify directly that

$$\frac{b}{a} \quad \text{is the multiplicative inverse of} \quad \frac{a}{b},$$

and that leads immediately to an unambiguous understanding of what it means to divide c/d by a/b.

Usually not too much is said about the real numbers at the elementary school level. After rational arithmetic, the next big leap is to learn the rudi-

---

[7]Of course there have been notable historical exceptions, such as Srinivasa Ramanujan.

ments of algebra. This shift is a genuine epistemological jolt, for it introduces abstraction into the mix. Now we use letters to denote unknown numbers, and that is a genuinely new idea. I cannot tell you how many times I have met people at cocktail parties who said to me (when they learned that I was a mathematician) that "I was always good at math until it got to that stuff with the letters—you know, algebra and all that abstract malarkey." What my cocktail-party companions have been telling me is that they failed the mathematical maturity game. They got to a crucial segment of mathematical development and were unable to muster the will or the tenacity to understand it and use it.[8]

Traditionally (when I was a high school student), the sophomore year entailed the study of Euclidean geometry. And that was (as it should have been) a milestone in intellectual development. In my Euclidean geometry course we spent an entire year studying axioms, stating theorems, and proving them. This was very exciting for me—as I was a budding mathematician —and it was a great academic step up the ladder. It is a sad fact that the high-school curriculum today is somewhat different in this regard. It is a basic theorem of abstract logic that any (proved) theorem may be declared an axiom. The modern crop of Euclidean geometry texts takes advantage of that result by introducing in each chapter twenty new axioms. In this way the book can avoid proving anything; the author knows that these results are true—indeed they have been proved elsewhere—so we can just adopt them as axioms. This is really very sad. For it means that we are turning a process of discovery and intellectual development into an empirical exercise of observation and discussion.

A next important segment in the learning curve is trigonometry. I am a pretty good mathematician, and I actually struggled as a student with several aspects of basic trigonometry. In trigonometry there are proofs that involve Euclidean geometry and analysis as well. The relationships among the trigonometric functions are subtle and sometimes even deep. To write down the equations for rotation of coordinates through an angle θ is rather tricky. Trigonometry is not for the timid. It is a stepping stone in the development of mathematical maturity.

And so the saga goes. Somewhere along this course of development the student is introduced to the concept of "function." Along with "set," this is one of the big ideas of twentieth-century mathematics. Every idea in every

---

[8]Distinguished mathematician Mary Ellen Rudin has always found this cocktail party repartee to be annoying. She says that she can just imagine herself saying to an English major, "Yeah, I was always good at *A*, *B*, *C*, but when it got to *D*, *E*, *F* and those other letters I fell apart."

modern mathematics tract and every modern mathematics research paper is formulated in the language of sets and functions. Functions are deep and subtle and very important. Only somebody with real mathematical talent can master the idea of function. Only someone who is mathematically maturing apace can actually use the language of functions to formulate mathematical ideas.

One of the reasons that the geniuses who invented calculus (Fermat and Newton and Leibniz) occasionally stumbled is that they did not have a clear concept of what a function is. It is a remarkable historical fact that the modern definition of function (as a set of ordered pairs that satisfies certain properties) was first formulated by Goursat in 1926 (see [KRA3]). There was much discourse in the time period 1850–1925 about what a function actually is, because this was the time in our history when many examples of pathological functions—for example the Weierstrass nowhere differentiable function—were created. Ralph Boas was a graduate student at Harvard in the early 1930s, and he relates (see [BOA]) that the professors at Harvard in those days could not agree on whether the (now famous) entity

$$ f(x) = \begin{cases} 0 & \text{if} \quad x \leq 0 \\ e^{-1/x^2} & \text{if} \quad x > 0 \end{cases} $$

is actually a function.[9]

The point here is a simple but fundamental one: mathematical maturity and mathematical inquisitiveness go hand-in-hand. One cannot flourish without the other. Each fuels the other. They develop side by side.

5.9 Mathematical Maturity is Not the Same as Knowledge

When I was young my father frequently spoke of "educated idiots." I am not sure that he quite knew what he meant by this, but it was his way of looking down on people who had a lot of schooling but not a lot of sense.

Just so, one can go to school and garner a good deal of knowledge without understanding anything. Many rather shallow people do just that. And they leave college wondering why they spent four years there and what good it did them. Such a person is not a candidate for mathematical maturity, or for any other kind of intellectual maturity. I have already made a case that

---

[9]Most of us encounter this function as an example of a C^∞ function that is not real analytic.

mathematical maturity must be built on knowledge, but it is not isomorphic to knowledge. They are quite distinct entities.

One can know all the rules for differentiation and integration and not understand the first thing about calculus. One can be able to recite every Galois group of any basic field extension of the rationals yet not know the first thing about abstract algebra. Speaking more generally, one can know a lot of facts and *not* know any mathematics.

Knowing mathematics leads to understanding mathematics. Understanding mathematics means being able to manipulate the ideas of mathematics— to build on the ideas you have in order to be able to reach out to new ideas. *That* is mathematical maturity.

Mathematical maturity is an ability to deal with, and a fundamental comfort with, abstraction. It is a way to take the long view with quite sophisticated ideas. It is a way of life. Gathering knowledge is just an idle distraction for dilettantes. It is neither productive nor inspiring.

5.10   Critical Thinking Skills

My father taught me many things, but the one that sticks out in my mind is that he taught me to say, "Now why in the heck should that be true?" And that is a useful attitude to have as a mathematician.

In mathematics there is no received wisdom. Everything is formulated axiomatically and proved rigorously. Any set of ideas that does not conform to the classical Euclidean paradigm is immediately rejected by the mathematical community. No, we do not take a vote[10] or have a discussion of the matter. For there is no need. By becoming mathematicians we have in effect taken the pledge: We do not believe with certainty what has not been proved.

There are well-known conjectures that are widely believed to be true: the Riemann hypothesis, the Goldbach conjecture, and so forth. But, as Branko Grünbaum once said, "An educated guess is still just a guess." It is mathematical proof that settles the argument once and for all. That is what makes us different from lawyers, from physicists, from engineers, and from philosophers. We have a failsafe means of testing our ideas *that is universally accepted in the profession*. There are other means to lend credence to a mathematical idea—key examples, computer calculation, model-building,

---

[10]Curiously, it was the habit of the Italian algebraic geometers in the 1930s to decide whether a new theorem was true by taking a vote. This habit was a product of certain personal ill will, and a nasty habit of secrecy about methodology and proofs.

reasoning by analogy—but the proof is the final arbiter. And we can take some comfort in that fact.

This view of the world sets us apart from virtually all other professionals.

- Literary critics make their points by analogy, by describing a topic from different points of view, by using language skillfully and persuasively. Literary critics do not often claim that they are offering proof of anything. They are instead staking out a position.

- Physicists make their points by experimentation, by examining "toy" versions of their problems, by doing seat-of-the-pants calculations, by numerical approximation, and by physical argumentation. Physicists may use the word "proof," but often that proof is a convenient form of discourse. It is not the formalism of mathematics.

- Chemists discover facts by way of hard-nosed experimentation. They want to measure things, and then determine what those measurements mean. Chemists certainly do not prove anything. They instead confirm physical reality.

To be fair, there *are* proofs in physics, but they play a different role than proofs in mathematics. As an example, we can and do prove Kepler's laws just using the hypotheses of Newton's law of gravity and Newton's laws of motion. It was Isaac Newton who first did this, to the astonishment of his scientific contemporaries. But this is not a theorem in the sense of a mathematical theorem. Mother Nature established Kepler's laws, Kepler himself confirmed them, and what Newton proved is that his laws of gravitation constitute a good theory, and give us a much better physical understanding than what we had before the time of Newton.

- Lawyers make their case by offering evidence. For a criminal case the evidence is supposed to establish a fact or contention "beyond a reasonable doubt." For a civil case it is "the preponderance of evidence shows." Lawyers may use the word "proof" in their presentation, but it is not mathematical proof to which they allude. Much of what a lawyer does is to argue about the meanings of words. Mathematicians never have that problem. The real world of justice and jurisprudence has no formal definitions and no axioms. Often lawyers are dealing with social matters that we agree on by custom or consensus. Of course the (written) law is very detailed and very precise, but it is not governed by strict rules of logic. A lawyer would never establish anything by induction. A lawyer would never cite *modus ponendo ponens*, although lawyers are as fond of Latin phrases as we are.

- Physicians are in the habit of treating patients. Their job is to cure diseases and rectify injuries. They use inference to learn from their past experience. They use logic to come up with diagnoses. But they do not prove theorems. They do not have any axioms (except perhaps "Do no harm."). Their few definitions are constantly subject to change.

- Most engineers have the attitude that their job is to get things done. They are not interested in abstract theory or pretty reasoning. If there is a mandate to develop an automobile that can get 40 miles per gallon in the city, then the engineer's job is to produce one. This need is how hybrid cars came about. If the goal is to develop a global positioning system (i.e., GPS—a navigation system) for cars, then the engineer's job is to develop one.

- A biologist endeavors to understand how living things work. This study involves a great deal of observation and organization of data. It often involves statistical reasoning. It does not involve theorems or proofs. It certainly does not involve axioms. There are no axioms for how a mammal works, or how a reptile works. There are collections of facts. That is *not* mathematics.

- A geophysicist once asked me whether mathematicians value originality and ideas. Without missing a beat, I said, "Sure. Originality and ideas are our meat and potatoes. That's what we live for." "Well," said the geophysicist, "in our subject we have no time for those guys with ideas and originality. They just confuse us. What is important in geophysics is organizing data." So draw your own conclusions. These folks are not proving theorems. They have no axioms or definitions. Their concerns are more pragmatic and empirical.

Part of the reason that non-mathematicians have a hard time understanding us, and understanding what we do, is that they do not share our values and they are relatively unacquainted with our methodology. The average person has no idea what it means to formulate a theorem or to prove a theorem. Most people think—and unfortunately their experience in K–12 school serves only to reinforce this misinformation—that mathematics is a done deal. Archimedes and Newton and perhaps Euler nailed it all down. There are no more questions to ask, no mountains to climb, no great discoveries to make. And it is quite difficult to explain to such people that they are misinformed. It is like a South African trying to explain to someone that there was never any *apartheid*; it was all just a social misunderstanding.

5.11   Ideas of Piaget

Jean Piaget (1896–1980) was a profound and influential developmental psychologist and philosopher. He has been called "the great pioneer of the constructivist theory of knowing." His ideas about education and teaching are still influential today.

Piaget conducted detailed studies of child learning, and of how children acquire knowledge. There is not space here to treat all these ideas. We would like instead to discuss briefly Piaget's *four stages of learning development*:

1. **The Sensorimotor Stage:** This stage is from birth to age 2. At this time, a child experiences the world through movement and sensory input. Children at this stage may be called egocentric, just because they cannot see the world from the point of view of another person. At the end of this stage the child segues from reaching out (tactilely) to experience new things to the beginnings of symbolic thinking.

2. **The Preoperational Stage:** This stage is from ages 2 to 7. Here the child is acquiring motor skills. At the beginning of this stage the child is still egocentric, but this property wanes over the 5-year period. The child is incapable of logical thinking or reasoning at this stage.

3. **The Concrete Operational Stage:** This stage is from ages 7 to 12. The child begins to reason logically, but only with the aid of practical tools. The child is no longer egocentric.

4. **The Formal Operational Stage:** This stage is from age 12 onward. Now the child begins to develop abstract thought processes and learns to think logically.

Piaget designed careful experiments to verify the details of each of these stages. We may take it that his model has some validity for child development. And it seems that it is also a model for the development of mathematical maturity. We can describe four stages as follows:

1. **The Naive Stage:** Students learns basic arithmetic skills. They do not worry about what math is good for or why they are learning these rote techniques. This sort of education is just routine learning of a Pavlovian sort.

2. **The Beginning of Awareness:** Students begin to manipulate ideas. They may begin to look at geometric figures and remark about their shapes and similarities.

3. The Dawn of Reason: Students begin to encounter abstract mathematical formulations, such as the Pythagorean theorem or the fundamental theorem of algebra. They may begin to wonder why these are true.

4. The Advent of Mathematical Thinking: Students start to see and experience proofs and to wonder about proving their own results.

There are many more stages of mathematical development beyond these four. We are only describing the evolution of a "mathematical child." It is easy to see how the precepts of Piaget have guided our thoughts.

Piaget is also noted for his theories to the effect that students learn a new idea by rebuilding that idea in their own minds. What we mathematicians more commonly say in this regard is that it is necessary for students to *internalize* a new idea so that they own it. In yet a third formulation, we might say that we do not really have control of an idea until we are ready to teach it, or to prove a new theorem using it. Every scholarly mathematician learns this last point of view from practical experience.

As an example, I did not really understand the proof of the Picard iteration scheme in ordinary differential equations until I had to teach it to sophomores. I did not understand what Plancherel's theorem meant until I had to teach it to a graduate class. I did not really understand the boundary regularity theory for elliptic partial differential equations until I created a course on the subject.

Likewise, I did not really understand differential forms until I had to use them in writing up my Ph.D. thesis. I did not understand what a subbasis for a topology was until I had to use the idea to understand spaces of real analytic functions. I did not really understand the invariance of curvature until I used it to prove an interesting theorem about the deformation of domains in complex space.

This is the space we occupy in the world of ideas, and this is how we ply our craft. Serge Lang would have said that he did not really understand a subject until he wrote a book about it. Most of us do not have the energy and insight of Serge Lang to carry out such a life-defining program.

We all find our own ways of internalizing ideas. Ed Dubinsky spent twenty-five years studying how college students internalize ideas. He ultimately devised the notion of teaching with the artificial intelligence software *ISETL*. He felt that, when students formulated mathematical ideas in *ISETL*, they were literally *forced* to rebuild the ideas in their own minds. Teaching at Purdue, he had some success with this approach. See [DUB1, DUB2, DUB3, DUB4, DUB5] for an exposition of some of Dubinsky's ideas.

We close by noting an interesting insight of Piaget. Asked to distinguish between logical concepts and non-logical concepts, Piaget replied that a concept is logical if it is completely reversible. That is to say, one can always reason from the concept back to its starting point. This is a remarkable insight. It validates why we hold the axiomatic method, which goes back to Euclid, so dear. Put in strong language: A concept is logical if you can prove it. Becoming mathematically mature amounts in large part to learning and understanding this assertion.

5.12 Reading and Thinking

It has been observed that the key things that a good teacher does are

- engage the students in the learning process

- pace the students

- teach the students to read

If you have wondered why someone trying to teach themselves the guitar generally tends not to make a lot of progress, and even after some time spent at the task is still clumsily banging out G, C, and D chords, it is because they cannot perform these three tasks for themselves. Everyone wants to be playing *Malagueña* or *Heartbreak Hotel* from day one. They have neither the discipline nor the tenacity nor the skill set to do the necessary finger exercises and drill so that they can build up to a suitable level of proficiency. In short, these students do not know how to pace themselves.

For my taste, it is the "reading" aspect of these three dicta that is the most interesting. If a student wants to learn calculus, why not just hand over a textbook and say, "Go forth and sin no more." The reason, of course, is that the student will be unable to read the book. That is all there is to it. There will be some exceptional students who can read the text. Most cannot. The teacher has to *show* the kids how to read the book. And learning to read mathematics is a critical step in developing mathematical maturity.

There is really nothing else like it. You learn that reading a math book is nothing like reading a comic strip. You are not going to do it while drinking wine and watching TV. You must concentrate, you must have a pencil in your hand, and you must frequently be stopping to cogitate and calculate and try things.

Reading mathematics develops in stages. You learn to read a calculus book by realizing that you must frequently pause to ponder, to calculate, to reason. This might occur two or three times per page. If you can read four pages per hour you are really tearing up the joint. When you get to upper-division work, the picture changes. Now you are reading perhaps Herstein's algebra book [HER] or Rudin's analysis book [RUD]. Now you have *long* pauses during which you must work out the reasoning for a tricky proof, or perhaps fill in the details of a nontrivial calculation. If you can read one page per hour you are doing well.

The reading of a graduate text is of the same nature, but even more demanding. Now the author expects more of the reader, and leaves out more detail. People used to say that, when Laplace said "obviously," then a couple of hours of hard work were in order. A graduate text should not be that tough, but it is certainly not for the casual reader. One or two hours per page is probably typical.

It is enlightening to recall here a passage from Descartes:

> But I will not stop to explain this in more detail, because I would deprive you of the pleasure of figuring it out yourself ...

Another great teacher in our midst.

Finally there is the reading of a research monograph. I still recall when I was a student reading Federer's *Geometric Measure Theory*. It took me a year to get through about 400 pages, and I was slugging away at it long into each night. There were many lacunae to fill in, and many missing calculations. But of course the process made me into a mathematician, and I am grateful for it.

Reading a medical school text or a law school text is dense and difficult. But it is really just a lot of facts. The process is *not* like reading an advanced math book. There is really no substitute for that experience. It demands that the reader go out on the edge and fight for the ideas. The payoff is terrific, but the process is daunting.

5.13   The Role of Writing

Writing is a key part of analytical thinking. The writing process entails taking your ideas apart and putting them back together again. It requires organization and planning and dissection. If you cannot write effectively and well, then that suggests that your thinking processes are not completely up to speed. It indicates that your understanding is not what it should be.

Learning how to write mathematics properly and incisively can easily be a part of the mathematical maturity process. And communicating your thoughts properly can also be part of that process. *Talking* to other people about mathematics is good for you. *Listening* to other people talk about mathematics is also good for you. Trying to figure out what they are saying and fitting it in with what is going on in your own head is a valuable yoga.

The process of planning a piece of writing—whether it be a research paper or an expository piece or a book—is a fulfilling and worthwhile activity. It forces you to re-think the subject, to put it in perspective, to analyze its component parts, and understand anew how they fit together. It creates a valuable synergy that you can capitalize on and use to push further in your understanding of the subject. It makes you stronger and wiser.

Actually *doing* the writing is even more valuable. For it is in effect explaining the ideas both to yourself and to your putative audience. If you are a good writer, you will be thinking about who your audience is and how they will be receiving your ideas. That in turn forces you to re-digest the ideas and recast them in a new argot. [As an instance, Andrew Wiles completely re-invented the theory (and the language) of elliptic functions in order to prove Fermat's last theorem [WIL].] And that is a useful exercise. You will see things in a new way and, no doubt, come up with new insights.

Part of being a scholar, part of being a mathematician, part of being a thinking and analyzing human being, should be writing. It puts you on the same plane as Descartes and Gödel—not a bad place to be. And it will teach you a lot about your life and your avocation. As Francis Bacon said,

> Reading maketh a full man, conference a ready man, and writing an exact man.

5.14 Tenacity and Delayed Gratification

Although we have eschewed the idea of probing personality traits that can lead to mathematical maturity, we shall here run the risk of contradicting ourselves and discuss two such attributes.

5.14.1 Tenacity

As an academic veteran of forty years' experience, I can say that being smart is a very powerful characteristic for a mathematician. But at least as important is perseverance or tenacity. Mathematics can easily be discouraging. At elementary levels, math is straightforward: you learn a new technique and can immediately sit down and apply it to a collection of exercises. There are few if any exceptions to the rules, and little ingenuity is required. Perhaps a little attention to detail is the main requirement.

But more advanced mathematics is different. Now you are dealing with ideas, and ideas are slippery creatures. They interact with each other in subtle ways. Ideas can appear to contradict each other, they can be resistive, and they can be uncooperative. If, in an exercise, you are asked to provide a proof or a counterexample, you may find that you are stuck for a goodly period of time. You may find that you simply do not have the right tools to solve the problem. You either do not know enough, or do not know the right things. This is why tenacity is important. You must have adequate faith in yourself to know that you can battle your way through the problem.

It is common today—at least in educational circles—to discuss math anxiety (see Section 4.1). A number of professionals have studied math anxiety to determine what it really is. In many cases they have determined that it is a combination of lack of confidence, math avoidance, and overall anxiety. Often math anxiety can be treated with a combination of relaxation techniques and remedial math review.

I might also humbly suggest that math anxiety is a byproduct of lack of tenacity—at least in the context of mathematics. Clearly a successful businessman, a chic interior decorator, or a prominent neurosurgeon will have plenty of tenacity in fields that they know and understand. But when it comes to the great unknown—in this case mathematics—these people are out at sea. They have no confidence, no foundation in experience or training, and (most importantly) do not have the will to press on. Such people are certainly *not* mathematically mature, and part of the reason is lack of perseverance.

A mathematician will often take several years to solve a problem. In fact it may take quite a few months just to master the problem, so that one can think about it deeply. Most people are not cut out for this. Jeff Ullman, a computer science professor at Stanford,[11] once said that you should never spend more than a year on anything. That may be fine and well in computer science, but it is not the path to success in mathematics. An attribute that is prized in the field of law is the ability to look at a complicated case and very quickly get to the heart of it and see the main point, and how it should be argued. Such a trait has almost no relevance in mathematical research. Of course we are all impressed by people who can see things quickly, but "quickly" is not the coin of the realm. It will not get you tenure in a good department, it will not make you famous as a mathematician, and it will not establish your legacy in the field. John von Neumann (1903–1957) was well known to be a lightning calculator and one who could learn almost anything very rapidly. But that is *not* why he is remembered. What we remember to-

---

[11] It should be noted that Stanford has one of the top three computer science programs in this country.

day is the theory of von Neumann algebras, the invention of Hilbert space, the invention (with Herman Goldstine (1913–2004)) of the stored program computer, the creation (with Oskar Morgenstern (1902–1977)) of game theory, and the mathematization of the theory of quantum mechanics.

5.14.2 Delayed Gratification

Not unrelated to the idea of tenacity is the property of being comfortable with delayed gratification. Rock stars and potentates tend to be arrogant and rapidly grow accustomed to getting whatever they want exactly when they want it—which is usually *right now*. Mathematicians are cut out for an entirely different quality of life. A mathematician often labors for years to achieve a particular goal. And, once it is achieved, you are prone to doubt that the time spent on it was worthwhile. After all, once you figure out how to do something (prove a theorem, or derive a formula, or generate a counterexample), then it all looks pretty obvious. Will anyone else be impressed, or even be interested?

Mathematicians spend their lives alternating between giddy elation and black despair. Many of us appear to be manic depressives, and some of us actually are. Delayed gratification is one of the main themes of our lives, and we must come to terms with it.

Certainly anyone striving for mathematical maturity must face delayed gratification head on. Elementary mathematics consists primarily of straightforward exercises—no delays there. But, once you get into theoretical mathematics, then you hit the wall. Now you are facing up to substantial ideas that really take time to master. Once you are challenged to generate your own proofs and counterexamples, you are frequently at odds, and often frustrated. It takes real time to solve some of these problems. And the process often consists of two steps forward and one step back. Many times what you learn today is that what you were thinking on the previous two days was wrong. But then your new take on the matter suddenly springs a leak, and you are back at square one.

The unifying theme for dealing with the need for tenacity and the need to deal with delayed gratification is self-confidence. You need to believe in your own abilities, and you need to believe that *you can actually do this work*. Not only that, but you must believe that *you yourself can do it on your own*. You are not going to get a tutor to help you. You are not going to find a solutions manual to tide you over. *You are going to do it.*

All of the truly successful mathematical scientists that I know—Fields Medalists and Nobel Laureates and Abel Prize winners—are supremely confident. This does not mean that they know they can solve any problem that comes their way. Far from it. They have faced as many frustrations and failures as the rest of us. But they have learned to deal with them. Perhaps it would be most accurate to say that these people know that they can face any mathematical situation (any problem, or any needed proof, or any needed counterexample) and learn from it and turn that new knowledge into something interesting and useful. And then perhaps write a paper and give some talks about it. Now *that* is mathematical maturity at its finest.

5.15   Types of Intelligence

Howard Earl Gardner is an American developmental psychologist who is John H. and Elisabeth A. Hobbs Professor of Cognition and Education at Harvard Graduate School of Education. He is best known for his theory of multiple intelligences. Gardner is famous for, among other things, specifying eight types or areas of intelligence. These are

1. linguistic

2. logic-mathematical

3. musical

4. spatial

5. bodily kinesthetic

6. interpersonal

7. intrapersonal

8. naturalistic

   It is interesting that items 2 and 4 on the list fall very naturally into the arena of mathematical talent, but none of the others do. This observation serves to emphasize that mathematical talent, or mathematical intelligence, is a *specialized* attribute. It is a small part of the overall mental terrain. It is well suited to certain types of activity, but definitely not for others. Linguistic intelligence may be useful in formulating your mathematical thoughts and for creating a written record (and subsequently for publication), but it is not central to the mathematical processes. I know many mathematicians who have a very highly developed musical aptitude, but it is not central to their mathematical activity. Of course interpersonal, intrapersonal, and kinesthetic intelligence have great intrinsic value, but they have nothing to do with mathematics.

Robert Jeffrey Sternberg is an American psychologist and psychometrician and Provost at Oklahoma State University. He was formerly the Dean of Arts and Sciences at Tufts University, IBM Professor of Psychology and Education at Yale University, and the President of the American Psychological Association. He has a triarchic theory of intelligence. So there are three areas or types of intelligence:

1. creativity

2. analytic

3. practical

This is a much more simple-minded view of the intelligence spectrum.[12] It is easy to see that items 1 and 2 have a lot to do with mathematical prowess, item 3 less so. We also note that, with Sternberg's view of the world, mathematical ability connects with 2/3 of the intelligence spectrum. But, with Gardner's view of the world, mathematical ability connects with just 1/4 of the intelligence spectrum.

Obviously these analyses of intelligence are somewhat subjective. The two lists that we see here are simply trees on which people can hang their hats. They are talking points when people attempt an analysis. They are not hard science.

One of the reasons that I am writing this book is to try to develop the idea of "mathematical maturity" from an intuitive feeling into a concrete, visceral set of properties or attributes that an individual may possess. Becoming a successful mathematics teacher consists in large part in understanding mathematical maturity and learning how to develop it.

5.16 Psychological Conditions

About twenty years ago, it became common credence that all genius has a common thread. Jackie Gleason (comedian), Willie Mays (baseball player), Albert Einstein (physicist), Isaac Stern (violinist), and Rudolf Nureyev (ballet dancer) all have a unifying trait. What is it? It is that they are all schizophrenic manic-depressives (see [LYE]).

This gives one pause for thought. Are we being told that, if one wants to become a world-class mathematician, then one must be mentally ill? Clearly one cannot simply *adopt* the properties of schizophrenia and/or manic depression. So is the die cast independent of any efforts that we may make on our own?

---

[12]It is amusing to note that people in the business refer to attribute 2 as "smarts" and to attribute 3 as "street smarts."

My view is that the property of genius is not uniquely defined, nor is it well defined. There are certainly some mathematical geniuses who exhibit at least some of the properties being described here. John Nash (a well-known schizophrenic) for example (see [NAS]) is a rather extreme example, but a valid one. Evariste Galois may have been another. But there are many others who seem to be quite normal.

One of the fascinating aspects of genius is that it is multi-faceted. Many geniuses are at first perceived to be eccentric cranks. It takes the world a good many years to appreciate what they have achieved, and what they are offering. As a simple instance, when Kurt Gödel first produced his incompleteness theorems, nobody noticed. It was only after vigorous campaigning, over many years, by John von Neumann, that the world began to appreciate what a significant contribution this was. Today we think of Gödel as one of the great mathematical minds of all time. Similarly, Einstein's relativity was at first viewed with great skepticism. When he was awarded the Nobel Prize (for his work on the photoelectric effect), he was advised *not* to speak about relativity at the Nobel ceremony. It was only many years later, when relativity was shown to have great predictive value in physics, that the full scope of Einstein's contribution was widely recognized and appreciated.

I know a good many Fields Medalists, Nobel Laureates, Abel Prize winners, National Medal of Science winners, and other luminaries. Each of these people is a dedicated, serious, committed scholar whose main goal in life is to contribute to their subject. These people all get up early and work hard all day to maximize the impact they can have on their field. To my amateur eye, they evince no evidence of pathology or mental illness. They are in many respects ordinary people, with spouses and children and houses and bills to pay. They simply have special gifts which enable them to accomplish more than the rest of us.

Just as with Asperger's syndrome (see Section 4.5), it would be a mistake to conflate mathematical talent with some medical condition. They surely influence each other, but they are not the same.

5.17 What are Our Students' Values? What are Our Students' Goals?

Our students are not like us. They have different backgrounds and different educations and different mindsets. They are, after all, frequently twenty or more years younger than we are. They are surrounded by, indeed drowning in, technology and information. They are *FaceBook*ed beyond recognition.

Communicating with people from a different generation is not a trivial matter. Of course one can just punt: march into the room, go to the lectern, and deliver a dry, canned lecture. That is not teaching. It is just going through the motions. *Teaching* means truly communicating. Getting inside their heads. Making them think. Making them react to what you are saying. As my Dean once said to me when I was Chair of the math department: "I want your kids, when they walk out of your classes, to be excited. I want them to be talking about the ideas. I want them to be pumped up."

Well, that is a tall order. It is easier to do this with political science or modern literature or philosophy than it is with mathematics. But it can be done. I have done it. You need to find a way to make them react. To see that what you are teaching them resonates with things that they care about.

One way to achieve this goal is to have class activities that will wake the students up. Eric Mazur's "Harvard peer instruction method" (see [MAZE]) of teaching physics is one tool worth pondering. Here is roughly how it works. The professor puts a pithy question on an overhead projector. In physics, such a question might be

> A physics professor in a boat lowers a rotating-type lawn sprinkler attached to a hose into a lake full of mercury. He attaches a pump to the other end of the hose and begins to pump out mercury. What does the sprinkler do?[13]

The students are instructed to discuss the question with the people on either side of them. After ten minutes, the instructor solicits answers from the audience.

It is easy to see that Mazur's is a lively teaching method. It is a way to get the students to think critically. It is a way to get them to learn to communicate. And it is a way to teach them some physics.

It would be easy to argue that the Harvard method is not a very good use of time. That one cannot get much physics covered with this methodology. Setting that question aside, I think that the more interesting question is, "How we can use this methodology in a mathematics class?" To be specific, how can we use it in a calculus class? I frankly do not know what the mathematical analogue of Richard Feynman's question might be. Mathematics has its own mode of discourse, and its own set of queries, and they are nothing like wondering about a rotating lawn sprinkler in a lake of mercury.

Ole Hald at U. C. Berkeley also has an interesting approach to getting the students turned on. He frequently teaches large (indeed, very large) calculus lectures. Let us say that today he is teaching the chain rule. Instead of just

---

[13] Some forms of this question have been attributed to Richard Feynman.

standing in front of the room and doing five examples—which is what most of us would do in the course of an hour—he does two brief examples and then he has the students do one. He actually has their problem prepared on little pieces of paper and has a gaggle of assistants to hand them out at the propitious moment. After the students attempt their problem, Ole says a few words about how it ought to be done. Then he does another example. And he concludes with the students doing one more example.

What a wonderfully original idea! When students listen to me lecture, they go home thinking that everything is crystal clear, and they have it down cold. But when they sit down a day or two later to do it themselves they cannot do anything. The advantage of Ole Hald's system is that he *empowers* the students. He shows them, on the spot, that *they* can do the problems. They leave the lecture knowing in a concrete way that they have learned, indeed have internalized, a new set of ideas. It is a wonderfully effective teaching technique, and it works. This comes fairly close to what my Dean was asking for: getting the students to acknowledge as they go out the door that they have learned something new and interesting.

In the short term, our students' goals are to graduate, to get that old sheepskin, and to find a good and useful place for themselves in society. We might hope that, in the long run, they want to become educated people so that they can live fulfilling lives. The point that I want to make is that it is *our job* to teach them that this insight is a worthy goal and something that they should be aiming for. It is part of teaching mathematical maturity, but it is also part of teaching intellectual maturity. Once again I will point out that one of our major roles as teachers is as paradigms for what an educated person should be. We *do* in fact live fulfilling lives. Most of us are quite happy with what we do and how we do it. We would like to pass some of that happiness on to our students.

5.18   Intuition vs. Rigor

Intuition is a vital part of serious, high-level thinking about almost any topic. Whether you are dealing with metaphysical epistemology or brain surgery or spectral sequences, it is difficult to navigate the minefield of ideas without some intuition of what is going on. Intuition is like a rough roadmap. It gets you headed in the right direction. It suggests some moves that you might try. It shows you some of the wrong roads and some of the right roads.[14]

---

[14] Although intuition is essential to creative thought, we must understand that it is *not* always reliable. It can certainly lead us astray. To be used effectively and well, the intuition must be developed and refined.

Intuition is certainly not a panacea. Nobody ever solved a deep problem using just intuition alone. But nobody ever made any intellectual progress without exercising at least *some* intuition. Once intuition gets you to the right general spot, then you must shift gears and apply *deep analytical powers* to make any further progress. But you would not be able to find that spot without some intuition.

From some perspectives, rigor is a counterpoint to intuition. It would be naïve to say that intuition is the product of right-brain activity while analysis is the product of left-brain activity. But it would not be completely wrong. I use my intuition all the time. As I get older, I find that my thoughts are more dominated by an intuitive approach to things (even very technical matters). But there is still analysis buried in my intuition. My intuition would be vapid and meaningless if it were not at least a bit analytical. After all, I am a mathematical analyst.

Part of maturing mathematically is learning to sort out intuition from rigor. Real tyros in mathematics are unable to separate the two. They will give an entirely intuitive, or graphic, argument for why the angle bisectors of an isosceles triangle must be of equal length and simply not realize that it is a heuristic argument. Or they might give a seat-of-the-pants reason why the function $1/(1 + x^2)$ has no minimum on the interval $[1, +\infty)$ but not realize that no actual *reason* has been given. Mathematical maturity means that you know the difference between jawboning and arguing mathematically. You can distinguish intuition from rigor. You can tell a proof from a plausibility argument.

When I teach calculus I never actually prove anything. I never even use the word "proof." My students have most likely never seen a proof, and have rarely heard the word. They hardly have any idea of what a proof is. So it would be beating them over the head to trot out a lot of proofs. What I do say to them is, "Here is an idea of why this is true." Or else, "Look at this diagram and you will understand the new fact that we are discussing." Or sometimes I say, "Take out your calculator and let us do a little calculation together. This will help us to understand the new paradigm that we have learned." Later on, when students take the department's "transitions course" (see Section 1.1), they will hear the word "proof" quite a lot. And they will learn to read proofs, to understand proofs, and to create their own proofs. But we must engage in this long and substantive process in steps.

Put in other words, I am not averse to making considerable use of intuition when I teach freshman calculus. As the curriculum moves forward, I phase out the intuition and replace it with rigor. When done carefully and with sensitivity to the students' viewpoint, it seems to work quite well. Students can be taught to downplay the role of intuition and replace it with strict rules of logic and rigor. Intuition is still there; it still plays a role. And the mature student will come to realize that rigorous argumentation can lead to a deeper level of intuition (and vice versa!). At the more advanced level, rigor plays the dominant role and intuition a more supportive role.

CHAPTER **6**

What is a Mathematician?

There are two kinds of results in mathematics: those that are obvious and those that are false.

Ron Getoor (mathematician)

For people with small horizons, every function is constant.

Oscar Bruno (mathematician)

It's only the false things that are nontrivial.

Michael Sharpe (mathematician)

Maturity is only a short break in adolescence.

Jules Feiffer (cartoonist)

Maturity is achieved when a person postpones immediate pleasures for long-term values.

Joshua Liebman (author)

Money does not buy happiness. I am now worth $50 million. But when I had only $48 million I was just as happy.

Arnold Schwarzenegger (body builder, actor, politician)

The world is everywhere dense with idiots.

L. F. S. (mathematician)

6.0 Chapter Overview

It seems clear that the role model for a student endeavoring to achieve mathematical maturity is the senior, successful mathematician. Such a person could be an academic professor with a vigorous research program and an international reputation. It could be a successful and innovative worker in the private sector—such as Robert Noyce, inventor of the memory chip and the microprocessor. It could be a leader at one of the many government

107

institutes—such as the National Security Agency (largest employer of math Ph.D.s in the world), or the Institute for Defense Analyses, or Oak Ridge National Laboratory, or Lawrence Berkeley Labs.

It is well to examine the attributes that contribute to the professional lives of successful mathematicians. How did they get to their current position of influence and success? What tribulations were met along the way? What can one learn from their career paths?

6.1   The Life of the Mathematician

I have said in other contexts that a mathematician is like a manic depressive. Working on mathematics problems is a tough life, often providing little more than discouragement and hopelessness. But that is the life of the single-combat-warrior researcher. You are out there on your own and you have nobody to rely on but yourself. It is lonely and it is tough, but the rewards (in the end) are great. And very satisfying. There is hardly any high so satisfying as that achieved from having proved a new theorem. Certainly the great classic book [HAR] gives a sense, from the perspective of a towering figure in modern mathematics, of the point being made here.

Not everyone is cut out for such a life. Not every human being can face such challenges. But that is the nature of mathematics. When one is a student in school, one must face the paradox of delayed gratification: You are frequently studying hard now for a payoff that will be a long time coming. When you are out there in the mathematics profession, then it is an even more challenging turf. You are betting your life and career against whatever problem you are currently working on.

6.2   Key Attributes of Mathematical Maturity

Sometimes it is useful to have a checklist of features of the topic being discussed. We provide one here. These are answers to the question "What is mathematical maturity?" It is more like a "to-do list" for those who want to be mathematically mature.

- Learn how to learn mathematics. Achieve the transition from rote memorization to learning for understanding.
- Learn to handle abstract ideas. Learn to learn abstract ideas without building up from concrete instances each time.
- Recognize mathematical patterns. Learn to internalize patterns and build on them. Learn to create your own patterns.

- Separate ideas from facts.

- Learn to draw upon already-learned or already-encountered skills and techniques in order to develop new problems and new techniques.

- Learn to read and analyze proofs. Learn to detect incorrect or sloppy proofs. Learn to create your own proofs.

- Move comfortably between the analytical (solving equations, analyzing inequalities, manipulating functions) and the visual or graphical (creating geometric representations, drawing graphs, implementing geometric visualization).

- Develop mathematical intuition by learning to abandon naïve assumptions and instead to analyze all hypotheses critically. Learn to draw on accumulated mathematical knowledge and experience to develop mathematical intuition and technique.

- Learn to develop links between analytical representations of ideas and geometrical representations of ideas.

- Learn to transfer mathematical techniques and mathematical insights to other areas such as computer science, biology, physics, and engineering.

- Learn to generalize from specific examples to broad concepts and insights.

- Learn to communicate mathematical ideas both orally and in writing. Learn to use standard terminology and standard notation. Determine how to teach that terminology and notation, together with the accompanying ideas.

- Learn to represent written and verbal problems as mathematical problems. Learn to introduce suitable variables, functions, and notation so as to represent a problem mathematically.

Notice that this list covers a broad array of skills. But it does *not* place an emphasis on facts or specific knowledge. It is more about a state of mind—a rather sophisticated skill set.

I still remember taking an algebraic topology course from Bill Browder at Princeton. It happened over and over again that we would be struggling for hours to complete a calculation to establish some isomorphism. Finally Bill would say to us, "Take the tensor product and then it's clear." And, after some thought, it usually was. Or talking to Rafy Coifman about harmonic analysis. More often than not he would say, "Dilate and look at the limit at infinity. Then it's clear." And, more often than not, this advice was dead on. When I worked at UCLA I would often ply Tony Martin with questions on the border of analysis and logic. His answer was invariably, "Well order the reals and then it's clear." This was an approach that I could never have thought of, and it usually worked. Nobel Laureate physicist Richard Feynman's favorite one-liner was, "Differentiate under the integral." He would give this as an answer whether he understood the question or not. And it was correct more often than not.

The point of the last paragraph is that the named mentors exhibited their maturity with cute one-liners, but these were in fact deep and important insights that they had mastered and I had not. I always had to give some thought to determine why their advice was correct, but it almost always was. I saw every time that they had seen to the core of the issue, and it was my job to battle my way to that same truth.

6.3   A Mathematician's Miscellany

John E. Littlewood was one of the more remarkable mathematicians of the twentieth century. He is remembered as G. H. Hardy's collaborator—they authored 100 papers together—but also as a powerful mathematician in his own right. He had a penetratring intellect, a strong appreciation of life, and a great sense of humor. His book [LIT] gives a sense of the kind of man he was.

Littlewood speaks in his book of the four rules that he and Hardy adhered to when conducting their seminal collaboration. Briefly, they are these:

(1) When one man wrote to the other, it was of no matter whether what he wrote was right or wrong. This gave them complete freedom to say whatever they liked.

(2) When one received a letter from the other, he was under no obligation to read it. Hardy was frequently noted to receive thick letters from Littlewood. He would toss them into a pile in the corner, saying that he might read them some day.

(3) Although it was all right for both of them to think about the same detail, it was better if they did not.

(4) [This is often said to be the most important of the axioms.] If one of them contributed very little or nothing to a given paper, it did not matter. Both their names would go on the paper, and they would both claim credit for the theorem.

These four "rules of life" are profound, and merit careful thought. They show both mathematical maturity and lifetime maturity. They exhibit a very long view of the collaborative process. Just as with rock-and-roll bands, the biggest bugaboo for mathematical collaborations is dissension over credit. Who contributed what? Who had the main ideas? Who kept the work going? Hardy and Littlewood realized that that was not the main point. Their chief desire was to keep working with each other. As Hardy makes clear in his *A Mathematician's Apology* [HAR], the most important fact of his life was his collaborations with Littlewood and with Ramanujan. The details of who proved which little result in which paper were of relatively minor importance. What Hardy valued was his relationship with these remarkable mathematicians.

Axioms 1 and 2 are particularly interesting. One thing that I enjoy about my own collaborators (and I have had 56 of them) is that we are completely comfortable with each other. When we get together to do mathematics, there is no pretense and no sense of one-upmanship. Our only goal is to create good mathematics. We have mathematical "rap sessions" together, and these are free and open, with no holds barred. Anyone can say anything that they like. It does not matter whether it is right or wrong. The point to is get the juices flowing. To get ideas out in the air so that we can examine them.

My Erdős number is 1, so I know a little bit about working with Paul Erdős. He liked to sit in a room with his collaborators and think. Frequently there were *very* long silences while people pondered the vicissitudes of life. But there were no rules and no formalities. Anyone could say or ask anything and expect a respectful, and we hope informative, reply. Sometimes Erdős would say something like "What is a compact set?" You might suppose that a senior mathematician like Erdős would *know* what a compact set was. But he just liked to hear the words. It got his thoughts moving. And the next day he might ask the very same question! But we respected each other, and we moved ahead, and we got the job done.

And, while it is great to have fellow mathematicians with whom to work, it is also great to have them at a bit of a distance. That way one can think one's own thoughts, without having to be constantly answerable to anyone else. When you are working by yourself you really *can* make all the mistakes you like, and plow up all the wrong paths you may choose. And learn in the process.

Littlewood also speaks of tenacity (in his collaboration with Mary Cartwright):

> Two rats fell into a can of milk. After swimming for a time one
> of them realized his hopeless fate and drowned. The other persisted,
> and at last the milk was turned to butter and he could get out.

In the first part of the war, Miss Cartwright and I got drawn into
van der Pol's equation. For something to do we went on and on at
the thing with no earthly prospect of "results": suddenly the entire
vista of the dramatic fine structure of solutions stared us in the face.

It is impossible to undervalue determination and doggedness in working on a mathematics problem. The healthiest thing (best for one's mental
health) is to work on several problems at once. That way, when one looks
particularly hopeless, you can set it aside and do something else for a while.
Coming back to the problem in six months or a year brings a fresh perspective and often new insights. I have many times spent twenty years on a
problem before solving it. But I did many other things in the interim.

Littlewood finishes his book [LIT] with a polemic on what it takes to be
a mathematician:

> There is much to be said for being a mathematician. To begin with,
> he has to be completely honest in his work, not from any superior
> morality, but because he simply cannot get away with a fake. It has
> been cruelly said of arts dons, especially in Oxford, that they believe
> there is a polemical answer to everything; nothing is really *true*, and
> in controversy the object is to prove your opponent a fool. We escape
> all this. Further, the arts man is always on duty as a great mind; if
> he drops a brick, as we say in England, it reverberates down the
> years. After an honest day's work a mathematician goes off duty.
> Mathematics is very hard work, and dons tend to be above average
> in health and vigor. Below a certain threshold a man cracks up; but
> above it, hard mental work *makes* for health and vigor (also—on
> much historical evidence throughout the ages—for longevity).

Littlewood tells a story of going through a period where he would dream
each night that he solved his research problem—and all the mathematical
details were there in the dream. But, when he woke in the morning, he
could remember none of it. He resolved to address this dilemma, and he put
a notebook and pencil by his bed. When, next night, he began to dream of
solving his problem, he forced himself to arouse and to write down all his
thoughts. In the morning he awoke with some anticipation, wanting to see
the details of his triumph. The notebook read

> Higamus bigamus men are polygamous. Hogamus bogamus wives
> are monogamous.

Littlewood said that you should be as intimate and comfortable with your
problem as you are with your tongue in your mouth. Your problem is not

something that you merely poke at, and nose around in once in a while. It is instead something that you live with and breathe with. The problem is like a spouse. It is something that gives meaning to your life, and succor, as well as frustration. You eat breakfast with your problem. You go for a run with your problem. You make love to your problem. You fight with your problem, and then you kiss and make up. Doing mathematics is a way of life.

Littlewood had the greatest admiration for Ramanujan. A man with no university education, Ramanujan had profound mathematical insights. Many of his greatest discoveries were remarkable formulas, which he produced virtually from whole cloth. In many instances he had no proof for the formula; he would just write it down.[1] He kept these formulas in a collection of notebooks, and these are still studied to this day by top-notch mathematical scholars. Many of these scholars have finally provided the proofs for some of Ramanujan's formulas.

---

[1] Ramanujan believed that these wonderful formulas were "handed to him" by God.

7

Is Mathematical Maturity for Everyone?

All you need for differentiation is a strong right arm and a weak mind.

Ron Getoor (mathematician)

A man whose mind has gone astray should study mathematics.

Francis Bacon (philosopher, writer)

Mathematicians, like cows in the dark, all look alike to me.

Abraham Flexner (educator)

The immature man wants to die nobly for a cause, while the mature man wants to live humanely for one.

Wilhelm Stekel (psychiatrist)

Those guys are talented; they're young; they're inexperienced. They need some maturity.

Rod Barnes (basketball coach)

All these prescriptions and descriptions about how to be a mathematician arose, inevitably, from my own attempts to become one. Nobody can tell you what mathematicians should do, and I am not completely sure I know what in fact they do—all I can really say is what I did . . . I taught, I wrote, and I talked mathematics for fifty years, and I am glad I did. I wanted to be a mathematician. I still do.

Paul Halmos (mathematician)

7.0   Chapter Overview

We tend to be a bit self-absorbed. We spend our days in mathematics departments (either at the university or elsewhere) thinking mathematical thoughts. So we are less than fully aware of the world around us.

As a result, we often do not have a full appreciation of the fact that there are other points of view in the world. Not everyone is passionate about mathematical maturity. How would a butcher or a baker or a literary critic or a chemist view mathematical maturity? Would any of them attach any value to the concept?

This chapter considers other points of view.

7.1   Who Needs Mathematical Maturity?

Mathematical maturity is an idea cooked up by mathematicians for the purpose of understanding the mathematical learning process. Clearly it is not for everyone. A person studying to be a gourmet chef probably has little need for mathematical maturity. A developing rock star has probably never given a thought to mathematical maturity.

But mathematical maturity is a paradigm for important modes of thought. And these models for reasoning have significance that extends beyond mathematics. There are many cutting-edge modern physicists, such as Ed Witten and Roger Penrose and Arthur Jaffe and Barry Simon and Frank Wilczek, who see physics axiomatically and mathematically. Understanding their stuff requires considerable mathematical maturity. Aspects of modern engineering, including microchip design and operating-system development and queueing theory, are extremely mathematical in both formulation and execution. There is no avoiding mathematical maturity if you want to participate in these enterprises.

It is also the case that mathematical maturity is a paradigm for what strict, rigorous, precise reasoning ought to be. It is a model for logic and for proof. The product of 2000 years of analysis and development, mathematical maturity is at the pinnacle of the evolution of human reasoning. It has intrinsic value, both for its purity and for its power. The way that the central processing unit in a computer carries out its tasks is a microcosm of mathematical maturity. The way that the bus on the back of your computer parcels out tasks is an implementation of mathematical maturity. The way that *Windows 7* manages your workflow is mathematical maturity in action.

If we want to understand ourselves better, and if we want to understand how to teach mathematics effectively and well, then we need to have a clear and decisive grip on mathematical maturity. We need to understand how to recognize it, and also how to nurture it and develop it. We need to know what parts of the teaching process are relevant to mathematical maturity and which parts are background noise. What is required here is *hard information*, *precision*, and *analysis*.

This book has been an attempt to provide some of the information adumbrated in the last paragraph. Together with history, background information, and motivation, we have endeavored to paint a picture of the entire mathematical maturation process. Using concrete examples from real life experiences, we have explained what mathematical maturity is, what it can be, and what it might become. Mathematical maturity is a part of our lives, and we must embrace it.

But we would do well to understand that there are other value systems, and other points of view. Someone studying to be a painter will be concerned with aesthetics and color values, and imagery. [It is true that M. C. Escher was profoundly inspired by mathematics (see [SCH]), but he was exceptional among painters.] Likewise, a developing real estate agent is interested in learning to identify and value properties. And how to present a property to a potential buyer so as to land a sale. Mathematics has little to do with it.

But that is what makes horse races. Everyone brings something different to the altar. Mathematical maturity is as important to us as making sales is to the real estate agent. Who can judge which is the more valuable? We need to respect and appreciate what everyone contributes to our lives; that is how we learn to live with each other.

7.2 The Role of Mathematical Maturity in Our World

It is a simple fact that most people are unaware of the concept of mathematical maturity—just as they are unaware of mathematics. Most people had a little math in school, did not enjoy it all that much, and were just as happy to put it behind them when school was over. Most people are *not* mathematical scientists, *not* quantitative analysts, and not likely to give these matters much thought.

This book is not intended for such people. It is obviously a book by a mathematician endeavoring to communicate with **(i)** other mathematicians, **(ii)** mathematics teachers, **(iiii)** those who might become mathematicians. For these folks, mathematical maturity is essential. It is our lifeblood. It is what we are about. Mathematical maturity is the nub of a mathematics education. The whole point of studying mathematics, from the modern perspective, is to learn to think like a mathematician.

It is essential for basic science, and for the advance of modern technology, that there be a nontrivial and identifiable segment of our society that consists of people who are conversant with mathematics and, yes, with mathematical maturity. So many modern advances—from GPS systems to cell phones to music CDs to operating systems—depend in an essential way on sophisticated mathematical theory. And one cannot come to grips with mathematical theory unless one has some mathematical maturity. It could be seat-of-the-pants mathematical maturity—garnered mainly from experience. But it has to be there.

In a world without mathematical maturity there would be a paucity of critical thinking skills. There would be no people who could do software verification, who could manipulate big ideas in a profound and effective fashion. There would be no cryptographers (in the modern sense of cryptography). There would be no theoretical computer scientists. There would be no image compression and no signal analysis. We would not have the SLIP software[1] for designing automobile bodies. We would not have the technology for tracking submarines on the other side of the world. We would not have cell phones.

The mathematically mature are our unsung heros. They keep alive a significant intellectual tradition, and one that is becoming ever more important. It will never be the *lingua franca* at Manhattan cocktail parties. But it will fuel the social advances that make those parties possible.

---

[1] Created by mathematician Bjorn Dahlberg of Washington University and a team from Volvo.

The Tree of
Mathematical Maturity

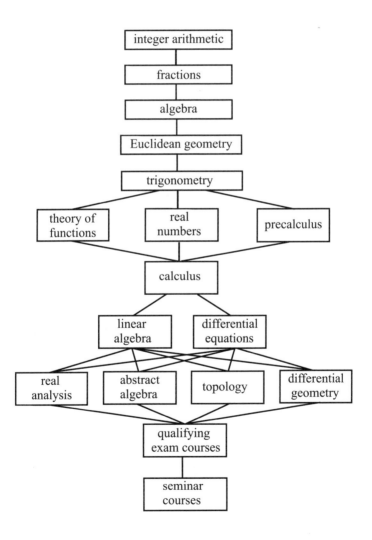

Etymology of the Word "Maturity"

The word "maturity" is derived from the Old French word *maturite* and from the Latin words *maturitas* (ripeness) and *maturus* (early, speedy, ripe)

Cognate words are the Dutch *matse*, English *matzo*, French *maturité*, German *Matura*, Italian *maturità*, Yiddish *matse*.

Bibliography

[AND]   D. Anderluh, Proposed math standards divide state's educators, *The Sacramento Bee*, October 26, 1997, p. A23.

[ASH]   N. Ashby, Relativity and the global positioning system, *Physics Today* 55(2002), 41–47.

[ALA]   D. Albers and G. Alexanderson, *Mathematical People*, Random House/Birkhäuser, Boston, 1985.

[BLS]   F. Black and M. Scholes, The pricing of options and corporate liabilities, *Journal of Political Economy* 81(1973), 637–654.

[BOA]   R. P. Boas, When is a C^∞ function real analytic?, *The Mathematical Intelligencer* 11(1989), 34–37.

[CAR]   L. Carleson, On convergence and growth of partial sums of Fourier series, *Acta Math.* 116(1966), 135–157.

[DAW]   J. P. D'Angelo and D. B. West, *Mathematical Thinking: Problem-Solving and Proofs*, 2nd ed., Prentice-Hall, Upper Saddle River, NJ, 1999.

[DEV]   K. Devlin, *The Math Gene: How Mathematical Thinking Evolved And Why Numbers Are Like Gossip*, Basic Books, New York, 2001.

[DOU]   R. G. Douglas, *Toward a Lean and Lively Calculus*, Mathematical Association of America, Washington, D.C., 1986.

[DUB1] E. Dubinsky, *ISETL*: A programming language for learning mathematics, *Comm. Pure and Appl. Math.* 48(1995), 1027–1051.

[DUB2] E. Dubinsky, J. Dautermann, U. Leron, and R. Zazkis, On learning fundamental concepts of group theory, *Educational Studies in Mathematics* 27(1994), 267–305.

[DUB3]   E. Dubinsky and W. Fenton, *Introduction to Discrete Mathematics with ISETL*, New York, Springer, 1996.

[DUB4]   E. Dubinsky and U. Leron, *Learning Abstract Algebra with ISETL*, Springer Verlag, New York, 1994.

[DUB5]   E. Dubinsky, K. Schwingendorf, and D. Mathews, *Applied Calculus, Concepts, and Computers*, 2nd ed., McGraw-Hill, New York, 1995.

[EC]   *Everybody Counts*,
          www.nap.edu/openbook.php?isbn=0309039770.

[FED]   H. Federer, *Geometric Measure Theory*, Springer-Verlag, New York, 1969.

[FEF]   C. Fefferman, The Bergman kernel and biholomorphic mappings of pseudoconvex domains, *Invent. Math.* 26(1974), 1-65.

[GKM]   E. A. Gavosto, S. G. Krantz, and W. McCallum, *Contemporary Issues in Mathematics Education*, MSRI Publications, vol. 36, Cambridge University Press, Cambridge, 1999.

[GAW]   Atul Gawande, *Complications: A Surgeon's Notes on an Imperfect Science*, Picador Press, New York, 2003.

[GLA]   M. Gladwell, *Outliers: The Story of Success*, Little, Brown, & Co., New York, 2008.

[GRA]   J. Gray, *Plato's Ghost: The Modernist Transformation of Mathematics*, Princeton University Press, Princeton, NJ, 2008.

[HAD]   M. Haddon, *The Curious Incident of the Dog in the Night-Time*, Vintage Books, New York, 2003.

[HAR]   G. H. Hardy, *A Mathematicians's Apology*, Cambridge University Press, Cambridge, 1940.

[HARR]   J. R. Harris, *The Nurture Assumption: Why Children Turn Out the Way They Do*, Free Press, New York, 2009.

[HER]   I. Herstein, *Topics in Algebra*, Xerox, Lexington, 1975.

[HAL]   D. Hughes Hallett, et al, *Calculus*, John Wiley and Sons, New York, 1992.

[JAC1]   A. Jackson, The math wars: California battles it out over math education reform (Part I), *Notices of the AMS* 44(1997), 695–702.

[JAC2]   A. Jackson, The math wars: California battles it out over math education reform (Part II), *Notices of the AMS* 44(1997), 817–823.

[KLR]   D. Klein and J. Rosen, Calculus reform—for the $millions, *Notices of the AMS* 44(1997), 1324–1325.

[KOW]   S. Kogelman and J. Warren, *Mind over Math: Put Yourself on the road to Success by Freeing Yourself from Math Anxiety*, McGraw-Hill, New York, 1979.

[KRA1]   S. G. Krantz, *How to Teach Mathematics*, 2nd ed., American Mathematical Society, Providence, RI, 1999.

[KRA2]   S. G. Krantz, *The Proof is in the Pudding: A Look at the Changing Nature of Mathematical Proof*, Springer, New York, 2011, to appear.

[KRA3]   S. G. Krantz, *The Elements of Advanced Mathematics*, 2nd ed., CRC Press, Boca Raton, FL, 2002.

[KRA4]   S. G. Krantz, What is several complex variables?, *Amer. Math. Monthly* 94(1987), 236-256.

[LET]   J. R. C. Leitzel and A. C. Tucker, Eds., *Assessing Calculus Reform Efforts*, Mathematical Association of America, Washington, D.C., 1994.

[LIT]   J. E. Littlewood, *Littlewood's Miscellany*, B. Bollobás ed., Cambridge University Press, Cambridge, 1986.

[LYE]   K. Lyen, Beautiful minds: Is there a link between genius and madness?, *SMA News* 34(2002), 3–7.

[MAZB]   B. Mazur, Mathematical Platonism and its opposites, `www.math.harvard.edu/~mazur/`.

[MAZE]   E. Mazur, *Peer Instruction: A User's Manual*, Benjamin Cummings, New York, 1996.

[MIL]   J. Milnor, On manifolds homeomorphic to the 7-sphere, *Ann. of Math.* 64(1956), 399–405.

[MOU]   D. Moursund, Math maturity,
        iae-pedia.org/Math_Maturity.

[NAS]   S. Nasar, *A Beautiful Mind*, 9th ed., Touchstone Books, New York, 2001.

[NCE]   National Commission on Excellence in Education, *A Nation at Risk: The Imperative for Educational Reform*, U. S. Government Printing Office, Washington, D.C., 1983.

[NOL]   William Nolen, *The Making of a Surgeon*, Mid List Press, Minneapolis, MN, 1999.

[PET]   Henry Petrowski, *To Engineer is Human: The Role of Failure in Successful Design*, Vintage Press, New York, 1992.

[RIC]   K. Richards, *Life*, Little, Brown & Co., New York, 2010.

[RIT]   J. F. Ritt, *Integration in Finite Terms*, Columbia University Press, New York, 1948.

[ROB]   A. W. Roberts, *Calculus: The Dynamics of Change*, Mathematical Association of America, Washington, D.C., 1995.

[ROI]   J. Roitman, Beyond the math wars, *Contemporary Issues in Mathematics Education*, MSRI Publications, vol. 36, Cambridge University Press, Cambridge, 1999, 123–134.

[ROS1] M. Rosenlicht, Liouville's theorem on functions with elementary integrals, *Pacific J. Math.* 24(1968), 153–161.

[ROS2] M. Rosenlicht, Integration in finite terms, *Amer. Math. Monthly* 79(1972), 963–972.

[ROSG] A. Rosenberg, et al, *Suggestions on the Teaching of College Mathematics*, Report of the Committee on the Undergraduate Program in Mathematics, Mathematical Association of America, Washington, D.C., 1972.

[RUD]   W. Rudin, *Principles of Mathematical Analysis*, 3rd ed., McGraw-Hill, New York, 1976.

[SCH]   D. Schattschneider, The mathematical side of M. C. Escher, *Notices of the AMS* 57(2010), 706–717.

[SIS]   J. Simons and D. Sullivan, Structured vector bundles define differential K-theory, *Quanta of maths*, Clay Math. Proc. 11(2010), Amer. Math. Soc., Providence, RI, 579–599.

[SPI]   M. Spivak, *Calculus*, 4th ed., Publish or Perish Press, Houston, TX, 2008.

[STK]   G. M. A. Stanic and J. Kilpatrick, Mathematics curriculum reform in the United States: A historical perspective, *Int. J. Educ. Res.* 17(1992), 407–417.

[STE1]   L. Steen, *Calculus for a New Century: A Pump, Not a Filter*, Mathematical Association of America, Washington, D.C., 1987.

[THU]   W. P. Thurston, On proof and progress in mathematics, *Bull. AMS* 30(1994), 161–177.

[TOB]   S. Tobias, *Overcoming Math Anxiety*, Norton, New York, 1978.

[TRE]   U. Treisman, Studying students studying calculus: a look at the lives of minority mathematics students in college, *The College Mathematics Journal*, 1992, 362–372.

[TUC]   T. W. Tucker, ed., *Priming the Calculus Pump: Innovations and Resources*, CPUM Subcommittee on Calculus Reform and the First Two Years, Mathematical Association of America, Washington, D.C., 1990.

[VER]   Abraham Verghese, *My Own Country: A Doctor's Story*, Vintage Press, New York, 1995.

[WIL]   A. Wiles, Modular elliptic curves and Fermat's last theorem, *Ann. of Math.* 141(1995), 443–551.

[WU1]   H. H. Wu, The mathematician and the mathematics education reform, *Notices of the AMS* 43(1996), 1531–1537.

[WU2]   H. H. Wu, The mathematics education reform: Why you should be concerned and what you can do, *Amer. Math. Monthly* 104(1997), 946–954.

Index

About the Author

Steven G. Krantz was born in San Francisco, California in 1951. He received the B.A. degree from the University of California at Santa Cruz in 1971 and the Ph.D. from Princeton University in 1974.

Krantz has taught at UCLA, Penn State, Princeton University, and Washington University in St. Louis. He served as Chair of the latter department for five years.

Krantz has published more than 60 books and more than 160 scholarly papers. He is the recipient of the Chauvenet Prize and the Beckenbach Book Award of the MAA. He has received the UCLA Alumni Foundation Distinguished Teaching Award and the Kemper Award. He has directed 18 Ph.D. theses and 9 Masters theses.